Advanced Water Injection for Low Permeability Reservoirs

Theory and Practice

Advanced Water Injection for Low Permeability Reservoirs

Theory and Practice

Ran Xinquan
Translated by Xu Fangfu, Gu Daihong

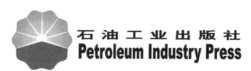

石油工业出版社
Petroleum Industry Press

AMSTERDAM • BOSTON • HEIDELBERG • LONDON
NEW YORK • OXFORD • PARIS • SAN DIEGO
SAN FRANCISCO • SINGAPORE • SYDNEY • TOKYO
Gulf Professional Publishing is an imprint of Elsevier

ELSEVIER

Gulf Professional Publishing is an imprint of Elsevier
225 Wyman Street, Waltham, MA 02451, USA
The Boulevard, Langford Lane, Kidlington, Oxford, OX5 1GB, UK

Notices
Knowledge and best practice in this field are constantly changing. As new research
and experience broaden our understanding, changes in research methods, professional
practices, or medical treatment may become necessary.

Practitioners and researchers must always rely on their own experience and
knowledge in evaluating and using any information, methods, compounds, or
experiments described herein. In using such information or methods they should be
mindful of their own safety and the safety of others, including parties for whom they
have a professional responsibility.

To the fullest extent of the law, neither the Publisher nor the authors, contributors,
or editors, assume any liability for any injury and/or damage to persons or property as
a matter of products liability, negligence or otherwise, or from any use or operation of
any methods, products, instructions, or ideas contained in the material herein.

Library of Congress Cataloging-in-Publication Data
A catalog record for this book is available from the Library of Congress.

British Library Cataloguing-in-Publication Data
A catalogue record for this book is available from the British Library.

ISBN: 978-0-12-397031-2

For information on all Gulf Professional Publishing
visit our website at http://store.elsevier.com

Printed and bound by CPI Group (UK) Ltd, Croydon, CR0 4YY

Working together
to grow libraries in
developing countries

ELSEVIER   Book Aid
            International

www.elsevier.com • www.bookaid.org

Contents

The reservoirs that have been discovered in the Ordos Basin, largely located in the Triassic and overlying formations, mainly consist of ultralow-permeability Mesozoic reservoirs characterized by arkose and lithic feldspathic sandstones developed through tectonic evolution and sedimentary diagenesis in the Basin. Initial development of the Basin targeted mainly the Jurrasic layers, with permeabilities over $10 \times 10^{-3}\ \mu m^2$. Large-scale development of the Ansai Oilfield between the late 1980s and the early 1990s was followed by quick and effective development of large ultralow-permeability Triassic oilfields, such as Jing'an, Xifeng, and Jiyuan. With permeabilities ranging between 0.5 and $2 \times 10^{-3}\ \mu m^2$, these reservoirs became the main development targets in the Changqing Oilfield. By the end of 2009, the proved reserves of these reservoirs had accounted for 78.8% of Changqing's total, and their outputs occupied 75.8% of Changqing's total production in 2009.

Featuring tight lithology, high flow resistance, low natural energy, low per-well output, and poor pressure conductivity, oil wells in ultralow-permeability reservoirs of the Ordos Basin have very low natural productivity, with quick decline and a primary recovery factor of only 8% to 10%. Therefore, artificial energy supply through waterflooding (and/or gas injection) to improve per-well productivity becomes the key to economical and effective development of these reservoirs, which, with special features, have their disadvantages, but also have their unique advantages. The key to their successful development is to make full use of their advantages, or rather, to overcome their disadvantages through developing core technologies fit for their features so that they can be put into cost-effective production.

Developed through years of practice, advanced water injection is one of the core technologies to tackle the unique features of oilfields in the Ordos Basin, such as ultralow permeability, high flow resistivity, low natural energy, low pressure conductivity, and low per-well productivity. It has proven to be a valuable technique to effectively improve well productivity.

For effective development of these oilfields, we started our practice on the basis of deep studies of the low-permeability reservoirs, carried out major research projects aiming at improving per-well output, ultimate recovery, and development efficiency, and experimented with advanced injection, well-group development, pilot waterflooded development, and industrial development. Five years of research and practice have made breakthroughs in tackling the bottleneck in developing ultralow-permeability oilfields, producing very good results.

During the large-scale development of Jing'an and other oilfields from 1997 to 2000, the strategy of synchronous injection was used to maintain reservoir energy, improve per-well output, and slow down output decline, with results much better than in Ansai. Then on the basis of production through delayed and synchronous injection, we put forward the theory of advanced injection and widely used it in Changqing.

To develop the theory, we did both theoretical studies and lab experiments between 2001 and 2007 to explore new theories and techniques fit for ultralow-permeability oilfields. As a result, the theory for advanced injection became systematic and complete. Taking into consideration the threshold pressure gradient and medium deformation and through nonlinear flow pattern studies, numerical simulation, lab experiments, and field tests, the theory analyzes the patterns of stress-sensitivity factors and threshold pressure gradients and studies the effects of medium deformation on petrophysical properties. Equipped with this theory, we developed a complete set of techniques centered on establishing an effective pressure displacement system through advanced injection. At the same time, auxiliary techniques for advanced injection were also investigated and developed, such as well pattern, well spacing, injection timing, injection rate, and reservoir pressure maintenance.

However, some practical problems, such as the unknown geological conditions and the backward ground facilities, still affected the practice of advanced injection. For this reason, we focused our efforts from 2008 to 2010 on developing techniques for quick evaluation of ultralow-permeability reservoirs and their productivity, so as to identify the oil/gas-rich zones and minimize the risks of drilling low PI wells and dry wells, thus providing a guide to the development strategy and improving the development efficiency. We also developed the mobile water injection skid, the smart flow-regulating valve complex, and the integrated digital skid pressurizer. At the same time, a new management system for advanced injection was established and practiced, such as planning of ground facilities in advance and nodal control. The practice of advanced injection is far more than an isolated technical job, but a systematic project, which can be summarized as "Three In-advances" and "Three Priorities." "Three In-advances" means predicting the development scale in advance, building a water injection system in advance, and constructing a water supply system in advance. "Three Priorities" means priorities should be given to water injector drilling, water injection pipeline construction, and investment in water injection before commissioning. With all these elements, the systematic water injection project acquires the features of "quick evaluation to set up its basis, overall planning to define its scale, innovative facilities to facilitate its practice and nodal control to promote its efficiency." Widely used in ultralow-permeability reservoirs such as Chang-6 and Chang-8, which are widely distributed across Huaqing, Jiyuan, Heshui, Wuqi, Huanbei, and Hujianshan, the advanced injection technique has resulted in very good development.

The central concept of advanced injection is to establish an effective pressure displacement system to reduce the formation damage caused by reservoir pressure drop, prevent the changes of physical properties of the oil in place, and improve the relative permeabilities of oil phases, so as to improve well productivity and oil recovery from the ultralow-permeability reservoirs. Our practice shows that the use of advanced injection can raise the per-well output by 20–30% and can effectively slow down output depletion.

This book aims to summarize the theory of advanced injection and its use in oilfield development in Changqing, so as to share our experience in developing these oilfields provide support for theoretical studies, and set a foundation for further studies of techniques for the development of ultralow-permeability oilfields.

Chapter 1

Features of Ultralow-Permeability Reservoirs in the Ordos Basin

Advanced Water Injection for Low Permeability Reservoirs.

1.1 GEOLOGICAL FEATURES

1.1.1 Structural Features

The second largest sedimentary basin in China, Ordos is a major cratonic-superimposed basin through long-term stable development over geological times, covering an area around $37 \times 104 \text{ km}^2$ across five provinces, i.e., Shaanxi, Gansu, Ningxia, Inner Mongolia, and Shanxi. Its western margin lies in the junction between the Sino-Pacific tectonic domain in the east and the Techyan tectonic domain in the west. Its southern margin is close to the junction between the two giant geological units of North and South China. Its southwest margin is bounded by a deep fault, neighbored by the Qilian and Qinling fold systems. Its northwest margin is adjacent to the Alashan block, and the northern part is joined to the Inner Mongolian axis in the form of an island arc. The central part of the Basin lies in the west of the North-China platform in geological processes, and is also a part of the Sino-Korean paraplatform. Though it has experienced quite a few tectonic movements, the central part of Ordos lacks internal structures because these movements usually take place in the form of vertical lifting and subsidence of the platform as a whole. Its present structure is just a large westward-dipping monocline with a dip angle less than $1°$ and a gradient of only $6 \sim 8$ m/km.

With reference to its current structural forms, the Basin can be divided into six first-order tectonic units, i.e., the Yimeng uplift, the Weibei uplift, the Jinxi scratch-fold belt, the Yi-Shaanxi slope, the Tianhuan syncline, and the thrust belt in the Western margin. The Yi-Shaanxi slope, containing a simple underground structure in the form of a large westward-dipping monoclinic with a dip angle less than 1°, developed a single type of reservoir, and therefore, became the mostly explored region with the most discoveries, including the upper and lower Palaeozoic gasfields and major oilfields with hundred million tons of oil, such as Ansai and Jing'an. This slope has thus become the main target of exploration and production in the Basin.

Decades of exploration and development, especially in-depth studies by generations of geologists, provide a general knowledge of distribution patterns of formations, structures, depositions, and resources within the Basin. In terms of hydrocarbon resource distribution, it is horizontally characterized by oil in the south and gas in the north, and vertically by gas in the lower formations and oil in the upper layers. According to regional tectonic evolution and sedimentary characteristics, the Basin has gone through five development phases: the aulacogen basin dominated by shallow marine clastic and carbonate rocks in the middle and late Proterozoic; the composite cratonic sag basin in which epicontinental carbonates are deposited in the early Paleozoic; the composite cratonic sag basin where coastal carbonates gradually transformed to clastic platform from the late Paleozoic to the middle Triassic; the sag basin where the large inland lakes and rivers deposit between the late Triassic and the Cretaceous; and the peripheral rift basin composed of inland river-and-lake filled rifts in the Cenozoic. These prototype basins, controlled by different tectonic movements, have both independent and interrelated evolution histories.

1.1.1.1 Stress-Field Distribution

An analysis of tectonic deformation traces and regional tectonic deformation characteristics within and around the Ordos Basin reveals that, after the sedimentation of the Triassic formations, the Mesozoic-Cenozoic tectonic stress field in the area can be divided into four stages, each with characteristics discussed as follows.

The Indosinian Period

Triassic Indosinian tectonic movements in the Ordos Basin produced an angular unconformity between the Jurassic and the Triassic, and parallel unconformity in other formations. The underlying layers, from the Carboniferous down to the Triassic, are linked in the form of conformity. According to the conjugate joints widely developed in the Carboniferous-Triassic strata along the Shanxi-Shaanxi boundary and the principal stress occurrence resulting from analysis of joints and microscopic rock fabrics within the Basin, the maximum principal stress axis in the Indosinian tectonic stress field is $10° \angle 2°$, while the minimum is $100° \angle 3°$. The middle principal stress occurrence is near vertical, a stress field with horizontally near north-south compression. An acoustic emission (AE) testing of the rocks from the Yanchang Group shows an average maximum effective principal stress of 93.5 MPa.

The Yanshanian Period

The Ordos Basin is dominated with mid-Yanshanian movements, with no obvious traces of early-Yanshanian movements. The large-scale uplifting of the eastern part in the late period of Yanshanian movements contributes to a structure of westward-tilting single dump in the Basin. Studies of the longitudinal bend folds, conjugate joints, initial sheet joints, shear zones, and microscopic rock fabrics in the Jurassic formations in the Basin and its surrounding areas show that the tectonic stress field during this geologic period has the maximum and minimum principal stress axes of $310° \angle 3°$ and 40°, respectively, while the middle principal stress axis is nearly vertical. The maximum principal stress axis calculated with statistics of the fracture occurrences observed at 45° along the Yanhe profile is $300° \sim 310° \angle 5°$, which reflects the distribution of the NW-SE horizontal compressional stress field in the Basin. AE testing on the Yanchang Group shows an average maximum effective stress of 83.5 MPa. The tectonic stress field serves as the controlling force for the development and distribution of fractures in the Basin in this period.

The Himalayan Period

As the northern part of the Indian-Australian plate moves north by east, and the Gangdese land mass moves quickly northwards and collides with the Qiangtang land mass, all land masses in western China move and collide northward, resulting in the N-NE horizontal compression of the tectonic stress field in the Ordos Basin during the Himalayan period. Studies of joints show a maximum principal stress axis of $30° \angle 2°$ and a minimum of 121°, respectively. The middle principal stress occurrence is again vertical, reflecting a distribution pattern of the NNE-SSW horizontal compression stress field over this period (Fig. 1-1). AE testing on Yanchang formations shows an average maximum effective stress of 62.5 MPa.

The Neotectonic Period

The magnitude and direction of the principal stress in the neotectonic stress field since the Pleistocene in the Ordos Basin can be measured with such methods as focus mechanism solution, hydraulic fracturing, and caliper caving. From the 133 focus mechanism solutions, 31 medium-to-small seismic data, and the practical data acquired by hydraulic fracturing and caliper caving in the Basin, the current stress field within this region shows the maximum principal stress in the NEE-SWE direction with horizontal compression and the minimum principal stress in the NNW-SSE direction with horizontal tension, averaging $70° \sim 80°$ in the NE direction (Fig. 1-1). The relatively tectonic stress values, usually below 10 MPa, have a great influence on the fracturing stimulation of sandstone reservoirs with ultralow permeabilities.

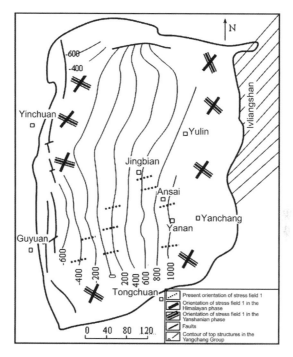

FIGURE 1-1   Distribution of Stress-field in the Ordos Basin at present time, Yanshannian and Himalayan periods.

1.1.2 Sedimentary Features

The Triassic Yanchang Group in the Ordos Basin is a large inland sag basin filled with river-lake facies sediments consisting of terrigenous clastics, controlled by southwest and northeast provenances. The Group contains 10 oil-bearing strata (numbered Chang-1 to Chang-10), whose sedimentary features reflect the evolution process from formation and development through to perishing of the lake basin.

The Chang-10 and Chang-9 members preserve the features of the formation-development period, when the basement subsided to form a lacustrine basin, with the sediments mainly consisting of fluvial facies and inshore shallow lake facies.

The Chang-8 and Chang-7 members were characterized by expansion of the basin. As the basin subsided and the coastline constantly expanded outwards, the lake basin became mature and reached a heyday, with its entire southern part covered with water. At that time, a large area of deep lake or half-deep lake facies developed in the present North Shaanxi. Therefore, the Chang-7 formation, mainly composed of gray to dark gray mudstones, became the major oil generator in the Basin.

The Chang-6 and Chang-4 + 5 members developed when the lacustrine basin began shrinking and deltas began to grow. While the Chang-6 formation was developing, the speed of basin subsidence was lower than that of sediment deposition. The active deposition produced large deltas and underwater fan-deltas around the basin, their distribution controlled by the flow direction of rivers that entered the lake and peripheral paleotopography. As a result, seven deltas grew in the flat northeast of the Basin from the lacustrine coastlines to the center, i.e., Yanchi, Dingbian, Wuqi, Zhijing, Ansai, Yan'an, and Niuwu. They appeared in groups and propagated to the center of the lake, causing the early phase of water regression and shrinking of the Basin. With a large area of the lake turning into swamps in the Chang-1 phase, the lake basin gradually became dry. Of all the Yanchang formations, Chang-9, Chang-7, and Chang-4 + 5 were three major subphases of lacustrine transgression. In particular, the Chang-7 subphase, when the lacustrine transgression reached its peak, developed the best source rocks deposited in the Mesozoic of the Basin.

Research shows that the Ordos Basin consists of two sedimentary systems. In the late Triassic Yanchang Group on the southwestern margin of the Basin, the sedimentation is characterized by multiprovenances, seriously deformed beds, steep basements, short transport distances, and a mixture of fine and coarse granules in the sediments, resulting in the subaqueous-fan and braid-river delta reservoirs of Chang-8 and Chang-6. On the other hand, in the same formation on the northeastern margin of the Basin, the sedimentation of Chang-6 is characterized by single provenance, slightly deformed beds, long transport distances and fine-grained sediments, producing a series of Chang-6 delta reservoirs, namely Yan'an, Ansai, Zhijing, Wuqi, and Yanding, from the east to the west.

Due to the joint influence of the sedimentary environment, dynamic conditions, transport distance, sedimentary topography, and diagenesis, the dominant reservoirs, i.e., Chang-8 and Chang-6 of the Triassic Yanchang Group in the Ordos Basin, are characterized by ultralow permeabilities. For example, in the Ansai Oilfield in the northeast of the Basin, the permeability of the Chang-6 reservoir in Wangyao is $1.9 \times 10^{-3} \, \mu m^2$, but in Yanhewan it is only $0.17 \times 10^{-3} \, \mu m^2$; in the Jing'an Oilfield, the permeability of the Chang-6 reservoir in Wuliwan is $1.81 \times 10^{-3} \, \mu m^2$, but it is only $0.59 \times 10^{-3} \, \mu m^2$ in the Wuqi Oilfield; and in the Xifeng Oilfield in the southwest of the Basin, the permeability of the Chang-6 reservoir is $1.77 \times 10^{-3} \, \mu m^2$ in the Baima block, but is only $0.11 \times 10^{-3} \, \mu m^2$ in Heshui.

1.1.3 Reservoir Features

1.1.3.1 Microfracture Features

Most ultralow-permeability oilfields that have been discovered and developed in China contain fractures. Generally speaking, the fractures developed under in-situ conditions in ultralow-permeability formations are mostly tiny fractures (or invalid fractures), and those parallel to or intersecting at small angles with the present direction of the principal stress field can be converted to effective fractures through artificial fracturing, thus improving the effective flow area and capacity of the reservoir. However, excessive pressure from fracturing or water injection may lead water into these fractures. In other words, fracturing in ultralow-permeability reservoirs has a dual effect: on the one hand, it can improve the water injection capacity, but on the other hand, it may lead to water fingering and thus too early waterflooding. Therefore, geological factors such as development features, distribution patterns, and effectiveness of fractures in ultralow-permeability reservoirs should not be neglected during oilfield development.

The fractures are well developed in the Ordos Basin under the stress along its margin, among which the Yansannion fractures in the WE direction are the most developed, followed by the NE fractures formed in the Himalayan period.

Fractures Observed in Outcrops

The observation stations for structural fractures in the Yanchang Group along the Yanhe profile involve members of Zhangjiatan, Qilicun, and Yongping. The geometrical features from different observation stations are slightly different (Fig. 1-2). Generally speaking, in Chang-6 most structural fractures along the Yanhe profile extend in the NEE-EW direction, some in the near SN (NNW-NNE) direction, and others in the NW direction. With their maturity closely related to lithology, the densities of fractures decreases from high to low in the order of muddy siltstones (11.7/m), siltstones (7.72/m), and fine-grained sandstones (4.64/m).

Fractures Observed in Core Samples

Statistics of cores from 31 wells of the Yanchang Group in the Ordos Triassic formations indicate that every well contains structural fractures with different degrees of maturity and usually high angles (Fig. 1-3). The fracture openings range from 0 to 0.3 mm, with the largest reaching 6 mm. The depths of these fractures vary greatly, 95% between 0 cm and 20 cm among which 50% are 0 cm∼5 cm. Less than 2% of all fractures have a depth of more than 50 cm. The fracture spacing in the cores presents an abnormal distribution, mainly 0 cm∼3 cm. The fractures are seriously filled, mostly with argillaceous sand and sometimes with calcite. So there is a low degree of communication between these fractures.

Most of the structural fractures observed are tension fractures, usually in a single direction, some in "X" directions and others in shear directions. Judged from the degrees of filling and mechanical features, most fractures in reservoirs Chang-6 to Chang-8 are ineffective.

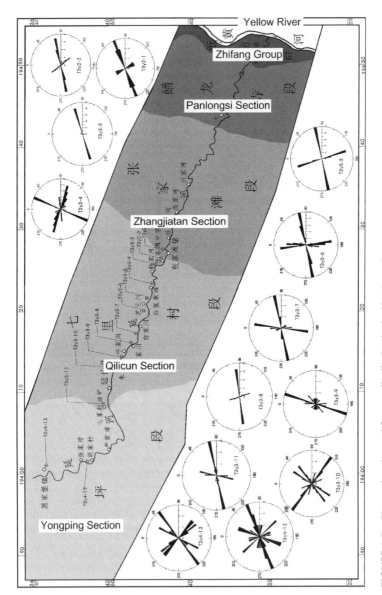

FIGURE 1-2 Observation sites and fracture distribution in the Yanchang Group.

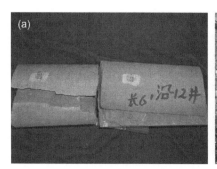

FIGURE 1-3   Pictures of cores containing fractures. (a) A high-angle natural fracture in Well Yan-12 in Chang-6 (interval 1987.45 m∼1987.76 m). (b) A zigzagging feldspar microfracture filled with bitumen in Well Zhuang-59–20 in Chang-8 (interval 2017.05 m∼2017.16 m).

Statistics show that the average apparent density of fractures in various sandstones is 1.01∼2.16 fractures/m. At the same time, the apparent densities of fractures of different lithologies decrease in the order of argillaceous siltstones, siltstones, and fine sandstones.

In spite of different features of development, the structural fractures of the Chang-6 to Chang-8 reservoirs have many things in common. Most fractures extend in the directions of nearly EW or nearly SN, with an opening of 0 mm∼0.3 mm, a depth of 0 cm∼20 cm, a spacing of 1 cm∼3 cm and a high angle. In addition, these structural fractures, mostly tension ones seriously filled and ineffective, may develop more often in siltstones than in mudstones.

1.1.3.2 Macroheterogeneity of Reservoirs

The macroheterogeneity of reservoirs involves the vertical and horizontal distribution of lithology, physical properties, oil saturation, and connectivity of sandbodies, mainly described in three aspects, i.e., interlayer heterogeneity, areal heterogeneity, and intralayer heterogeneity.

Interlayer Heterogeneity

Interlayer heterogeneity refers to the differences in the geological factors that control the concentration and flow of subsurface fluids between formations. An index of formation-scale reservoir description is a kind of macrostudy of the oil layers within a reservoir or a member of the interbedded sandstones and mudstones. As an internal cause of interlayer interference, waterflood efficiency, monolayer breakthrough, and distribution of the remaining oil during the waterflooded development, interlayer heterogeneity serves as the basis for selecting target layers for development and corresponding techniques for developing different layers. The study of interlayer heterogeneity, with oil layers as the appraisal units, uses the thickness of a single sandbody, the stratification coefficient, the permeability variation

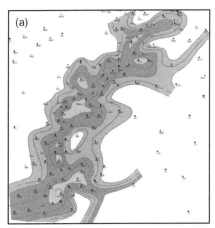
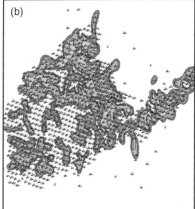

FIGURE 1-4   Distribution of mudstone barriers in typical areas. (a) Distribution of mudstone barriers in Chang-8. (b) Distribution of mudstone barriers in Chang-6.

coefficient, the heterogeneity coefficient of permeability, the permeability differential, and the distribution of barriers to describe the size, the flow quality, and the interlayer differences in a single sandbody.

The sand coefficients of the ultralow-permeability reservoirs in Ordos range from 0 to 57, mostly $12 \sim 40$, averaging 25.7. Their sand densities range from 0% to 97%, among which those of $0\% \sim 29\%$ account for 39.3%, while those of $30\% \sim 50\%$ account for 27.7% and those of $51\% \sim 100\%$ account for 43.0%. Most of the sand densities lie between 30% and 70%, averaging 44.7%.

Barriers abound in the ultralow-permeability Triassic reservoirs, with claystones, siltstones, silty mudstones, and calcareous siltstones as their lithologies. Containing no oil, these barriers have the function of isolating fluid flows. The relatively steadily distributed barriers are usually $4\,\mathrm{m} \sim 20\,\mathrm{m}$ thick, smaller in the main sandbody zone and larger in the side flanks (Fig. 1-4).

Areal Heterogeneity

Areal heterogeneity refers to the areal changes of the geological factors that control oil-layer distribution and influence the concentration and flow of subsurface fluids, including the geometric shape, size, and continuity of a sandbody, the microstructure of a reservoir, and the areal changes of porosities, permeabilities, thicknesses, and effective thicknesses within a sandbody.

The Geometric Shape, Size, and Continuity of a Sandbody and Its Communication with Others

There are four main geometric shapes of sandbodies in the Ordos Basin, i.e., the sheet shape, the ribbon shape, the potato shape, and the dendritic shape, with the ribbon-shaped sandbodies the most developed (Fig. 1-5).

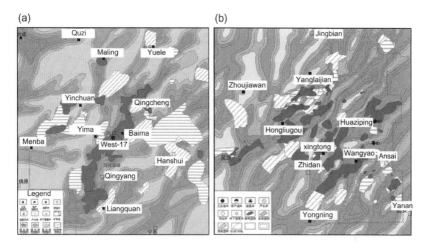

FIGURE 1-5 Distribution of sandbodies in the Ordos Basin. (a) Distribution of sandbodies in Chang-8, Longdong block. (b) Distribution of sandbodies in Chang-6, Longdong block.

The focus when studying the size and continuity of sandbodies is on their lateral continuity, with the width-thickness ratio, drilling ratio, and quantitative geological database as the main indexes to describe and predict such continuity.

From the viewpoint of origins, the connectivity of subsurface sandbodies falls into two types, i.e., structural connectivity and sedimentary connectivity. While the former is caused by faults or fractures, the latter is caused by vertical and horizontal connection of sandbodies, which may be expressed with such indexes as sand coordination number, connectivity and connectivity coefficient. The connectivity not only concerns the development well density and well pattern of waterflooding, but also affects the final recovery factor. An analysis of the vertical and horizontal connectivity profile of per-well grids indicates that there are three main types of sandbodies in terms of connectivity in the Chang-6~Chang-8 reservoirs in the Yanchang Triassic Group in Ordos FIGURE 1-6.

The multiboundary sandbodies. Sandbodies of different origins finger-cross with each other in the lateral direction, as the underwater distributary channel sandbodies and the river-mouth bar sands at the front of deltas in this block often do.

The multilayer sandbodies. Sandbodies of different origins are connected with each other in the vertical direction. This is a kind of superimposed connection of channel sandbodies caused by the migration of channel branches.

The isolated sandbodies. Sandbodies refer to those enclosed by mud or sandbodies isolated from other sandbodies by impermeable barriers.

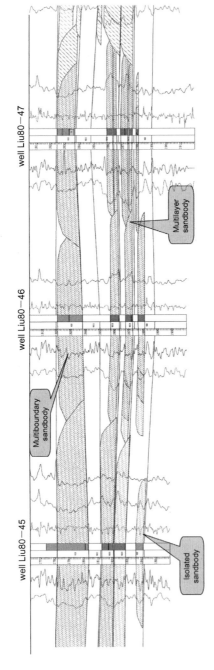

FIGURE 1-6 A cross-section of the Chang-6 reservoir in Wuliwan.

Characteristics of Physical Property Distribution

The horizontal distribution of porosities and permeabilities of the Chang-6 ~ Chang-8 reservoirs in the Yanchang Triassic Group in Ordos is characterized by obvious heterogeneity (Fig. 1-7 and 1-8). Generally speaking, the zones where there are thick sandbodies often have good physical properties and high oil saturations while those with thin sandbodies have opposite features. The areal heterogeneity of reservoirs is controlled by the degree of development and the microsedimentary facies of sandbodies. In terms of the microfacies, the sandbodies mainly consisting of underwater distributary channels at the front of deltas are often the thickest and have the best physical properties while the river-mouth sand bars and sheet-shaped bodies at the front of deltas follow. In terms of degree of development, reservoirs with continuous multilayer superimposed sandbodies have good physical properties and homogeneity while those with separated or gridded sandbodies are greatly heterogeneous and have poor physical properties.

Intralayer Heterogeneity

Intralayer heterogeneity refers to the comprehensive geological factors that control and influence the vertical flow and distribution of fluids within

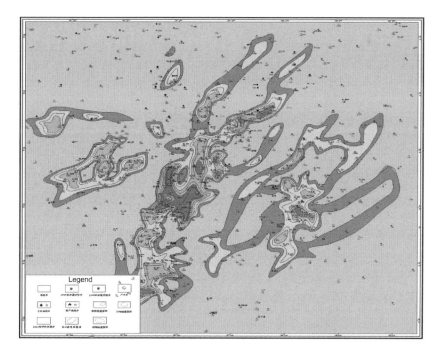

FIGURE 1-7   Permeability contour map of the Chang-6 reservoir in the Yanchang Group, Longdong block.

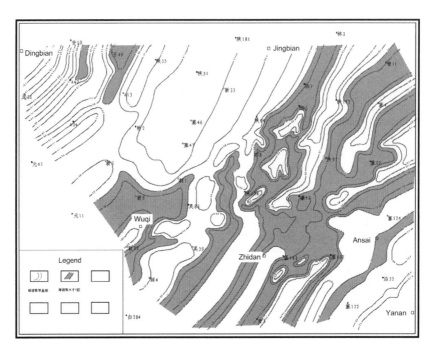

FIGURE 1-8 Permeability contour map of the Chang-6 reservoir in the Yanchang Group, North Shaanxi.

a single sand layer, mainly including particle rhythm features, bedding structures, permeability rhythms, ratios of vertical to horizontal permeabilities, permeability heterogeneity, and distribution frequencies and densities of muddy bands. In general, different permeability features involve different waterflooding modes, which cause differences in development results and distribution patterns of the remaining oil. This book will discuss the intralayer heterogeneity in the Ordos Basin focusing on the permeability variation coefficient, the heterogeneity coefficient of permeability, and the permeability differentials (i.e., the ratio of the maximum permeability to the minimum).

The minimum permeability of the Chang-8 reservoir in Ordos is $0.02 \times 10^{-3}\ \mu m^2$, and its maximum is $10.09 \times 10^{-3}\ \mu m^2$, with the variation coefficients between 0.52 and 85.03, the heterogeneity coefficients between 1.45 and 13.75, and the differentials between 2.27 and 1766 (Fig. 1-9). As for the Chang-6 reservoir, its variation coefficients range from 0.23 to 0.69, its permeability differentials from 3.13 to 329, and the heterogeneity coefficients from 1.76 to 22.0 (Fig. 1-10). The sedimentary rhythms are controlled by sedimentary facies, with the zone of river-mouth bars in inverse rhythms and the zone of distributary channels in positive or mixed rhythms.

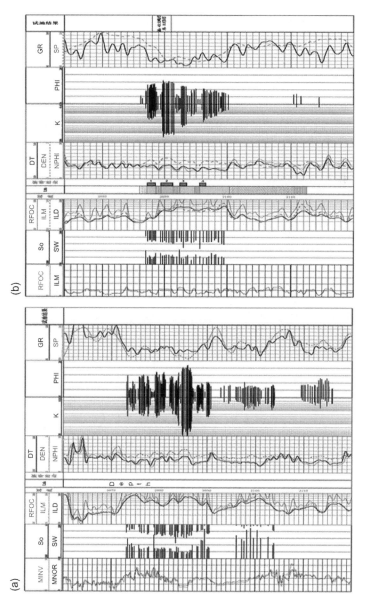

FIGURE 1-9 Relations between four physical properties in Chang-8, Longdong block. (a) Relations between four physical properties in Chang-8, Well Xi-119. (b) Relations between four physical properties in Chang-8, Well Xi-23.

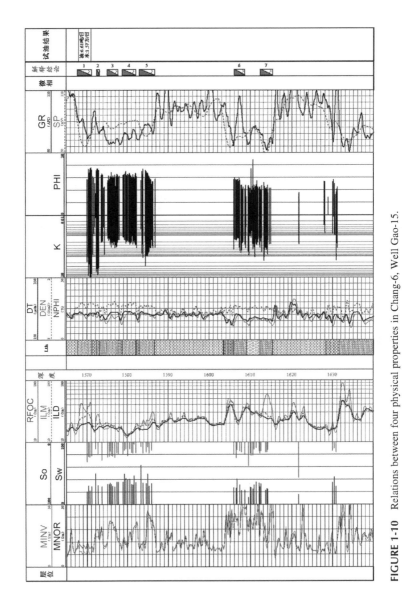

FIGURE 1-10   Relations between four physical properties in Chang-6, Well Gao-15.

The intralayer bands in the Chang-8 reservoir are tight formations of calcareous cementation cut by river courses and superimposed at the river bottom. Logging curves are characterized by a decline of sound wave values but an increase of resistance values. Because of extremely low permeabilities, these bands, usually 300 m $\sim$ 500 m long and 0.5 m $\sim$ 2 m thick, do not contain any oil. The degree of development and distribution of these bands are mainly influenced by hydrodynamic conditions and sediments at the time of deposition. The intralayer heterogeneity of Chang-8 is greatly influenced by sedimentary microfacies, declining from north to south. The features of intralayer bands in Chang-6 are similar to those in Chang-8, mostly calcareous tight formations. Few (0–2) of the bands develop in the river-mouth sand bars and distributary channel sandbodies, but more (3–5) develop in river flanks, distributary interchannels, and underwater natural levees.

To sum up, sedimentary facies are the main factors that influence the intralayer heterogeneity of formations in the Ordos Basin.

1.1.3.3 Microscopic Features of the Reservoirs

Pore Types

The Ordos Basin develops mainly primary and secondary pores, plus a small number of microfractures.

Primary pores Primary pores mainly refer to intergranular pores between clastic particles, including residual intergranular pores (Fig. 1-11) and miscellaneous micropores (Fig. 1-12).

Secondary pores Secondary pores include dissolution pores (Fig. 1-13) and intergranular dissolution pores (Fig. 1-14).

FIGURE 1-11 Development features of residual intergranular pores in Ordos Basin: (a) Residual intergranular pores at Well Xi-15 of Chang-8 reservoir at Longdong (2017.65 m), (b) Well-developed intergranular pores at Well Sai-263 of Chang-6 reservoir at North Shaanxi (1807.34 m)

FIGURE 1-12 Developmental features of miscellaneous micropores in Ordos (Chang-6, Well ZJ-54, 1813.24 m).

FIGURE 1-13 Developmental features of feldspar intragranular dissolution pores in Ordos. (a) Well-developed feldspar dissolution pores in Chang-6, Well Pan-26—44 (1907.65 m). (b) Well-developed dissolution feldspar in Chang-8, Well Xi-30 (2095.13 m).

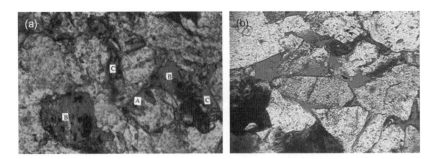

FIGURE 1-14 Developmental features of feldspar intergranular dissolution pores in Ordos. (a) Enlarged dissolution intergranular pores, moldic pores, and intergranular dissolution pores in Chang-6, Well ZJ-6 (1897.63 m). (b) Intergranular dissolution pores in Chang-8, Well Xi-163 (2114.45 m).

FIGURE 1-15   Developmental features of microfractures in Ordos. (a) Diagenetic fractures in Chang-6, Wells Zhuang-125−19 (1899.43 m). (b) Quartz fractures in Chang-8, Wells Zhuang-58−22 (2099.93 m).

Microfractures   Microfractures only develop in some sandstones. Most of these are interstratal fractures developing along planes rich in black mica or phytoclasts, and some are tension fractures with diagonal bedding. Microfractures are mostly tensile, with little fill. Some have dissolution pores on both sides (Fig. 1-15).

Structural Features of Pore Throats

An analysis of the data acquired through conventional mercury penetration, cast-slice imaging, and constant-velocity mercury penetration shows that pore throats in the ultralow-permeability reservoirs in Ordos have the following structural features:

- The pore throats are all slender, mostly small pores with microthroats.
- The throats are distributed in bimodal and skew patterns.
- The porosity and the permeability are positively correlated, mainly forming porous reservoirs, while the permeability and the displacing pressure, median pressure, and mean pressure are negatively correlated.
- When the permeabilities are less than 1×10^{-3} μm^2, the distribution of throat is centralized, leading to a high proportion of peak radiuses, and when the permeabilities are greater than 1×10^{-3} μm^2, the throat is more scattered, resulting in a lower proportion of peak radiuses.
- As the permeability increases, larger throats also increase considerably, are less centralized, and are more widely scattered.
- The reservoir quality is entirely controlled by throats from a microviewpoint.
- The microheterogeneity of reservoir is relatively strong, and larger throats make an important contribution to oil and water flows in reservoirs with ultralow permeability.
- All these features indicate a strong microscopic heterogeneity of reservoirs in the Ordos Basin (Table 1-1).

TABLE 1-1 Structural Parameters of Pores in Chang-6 to Chang-8 Triassic Reservoirs, Yanchang Group

Reservoir	Permeability $(10^{-3}\ \mu m^2)$	Average throat radius (μm)	Main throat radius (μm)	Isotrope factor	Relative sorting factor
Chang-6	0.17	0.46	0.51	0.37	0.60
	0.17	0.40	0.46	0.27	0.81
Chang-8	0.25	0.52	0.69	0.12	0.53
	0.38	0.60	0.77	0.53	0.32
	0.50	0.72	1.14	0.31	0.54
	0.94	0.98	1.80	0.09	0.71
	1.72	2.53	3.58	0.22	0.68
	3.14	2.70	3.54	0.35	0.50
	4.47	2.81	3.71	0.30	0.59
	6.40	2.75	3.72	0.31	0.64
	8.08	3.29	4.33	0.36	0.55
	13.25	3.56	4.51	0.35	0.61

1.1.3.4 Reservoir Wettability and Sensitivity

Reservoir Wettability

Our analysis of 226 cores from the Chang-8 to Chang-4 + 5 reservoirs in the Triassic Yanchang Group implies that Chang-4 + 5 is neutral to water-wet, Chang-6 is mainly neutral, and Chang-8 on the whole is neutral, but neutral to slightly oil wet in the Baima and Zhenbei blocks of the Xifeng Oilfield.

Reservoir Sensitivity

Reservoir sensitivity includes water sensitivity, acid sensitivity, salt sensitivity, alkali sensitivity, speed sensitivity, and stress sensitivity. While the first five are evaluated by lab tests of core flows, the stress sensitivity is evaluated through changes in the perm-plug permeabilities and porosities at different overburden pressures and compressibility factors of rocks. The stress sensitivity is an important feature of ultralow-permeability reservoirs, which will be elaborated on later.

Statistics of lab tests of 206 cores from the Chang-8 to Chang-4 + 5 reservoirs in the Yanchang Group show that its Triassic reservoirs have the features of weak speed sensitivity, weak water sensitivity, weak salt sensitivity, medium-to-weak acid sensitivity, and weak alkali sensitivity.

1.1.3.5 The Temperature-Pressure Systems

According to the statistics of the initial reservoir pressures, reservoir depths, and formation temperatures of the 24 developed blocks in the ultralow-permeability Triassic reservoirs in the Ordos Basin, the pressure factors range from 0.62 to 0.89, with an average of 0.74, indicating low pressure in them. Our studies also show that the differences between reservoir pressures and saturation pressures remain small, usually between 2.94 MPa and 6.38 MPa. The formation temperatures in these blocks range from 40 °C to 80 °C while the geothermal gradients range from 3.0 °C/100 m to 3.2 °C/100 m, indicating a normal temperature system.

1.1.3.6 Fluid Properties

Oil Properties

The crude oil in these reservoirs enjoys good physical properties, with the ground crude showing low specific gravities, low viscosities, low asphaltene contents, low freezing points, and low initial boiling points, and is sulfur-free and nonwaxy. The formation crude has viscosities between 1.48 and 4.3 mPa.s, averaging 2.39 mPa.s, densities from 0.693 to 0.8 g/ml, averaging 0.752 g/ml, and gas/oil ratios between 47.8 and 97.38 m^3/t, averaging 72.82 m^3/t (Table 1-2).

Formation Water Properties

The formation water in the reservoirs of Chang-4 + 5, Chang-6, and Chang-8 has the following properties: high Ca^{2+} content, high $K^+ + Na^+$ content, high Ba^{2+} content, and low HCO_3^- content, with little or no CO_3^{2-} and SO_4^{2-} ions. With salinities from 70 g/L up to 132 g/L, the formation water is typical of high salinity (Table 1-3).

Properties of Associated Gases

The nonhydrocarbon components of the associated gases in the Triassic reservoirs in Ordos have low CO_2 content but high nitrogen content. Compared with other reservoirs in the Basin, Chang-8, Chang-6, and Chang-4 + 5 contain associated gases with relatively higher contents of methane and heavy hydrocarbons, the latter between 30% and 60% (Table 1-4).

1.1.3.7 Saturation of Movable Fluids

Reservoir development not only pays attention to geologic characteristics and rules of fluid flow, but also tries to understand the amount of movable fluids in the reservoir. The fluid evaluation of 116 cores from three formations in the six oilfields in the Yanchang Triassic Group of the Ordos Basin through nuclear magnetic resonance shows that the NMR T_2 spectrum of most cores is bimodal. According to classification criteria observed both at home and abroad, Type-A reservoirs account for 33.6%, Type-B reservoirs

TABLE 1-2 Physical Properties of Oil in Place

Reservoir	Reservoir pressure (MPa)	Saturation pressure (MPa)	Coefficient of compressibility (10^{-4}/MPa)	Crude viscosity (mPa·s)	Gas/oil ratio (m^3/t)	Volume factor	Crude oil density (g/ml)	Dissolved factor (m^3/m^3/ MPa)	Relative gas density
Chang-4 + 5	10.52	8.50	9.30	4.30	47.80	1.330	0.800	6.95	1.144
Chang-6	10.24	8.75	9.89	1.83	78.70	1.212	0.693	7.43	1.143
Chang-8	15.22	10.80	12.47	1.48	97.38	1.284	0.746	7.65	1.141
Average	11.340	8.83	10.54	2.39	72.82	1.256	0.752	7.49	1.143

TABLE 1-3 Chemical Properties of Formation Water

Formation	$K^+ + Na^+$	Ca^{2+}	Mg^{2+}	Ba^{2+}	Cl^-	SO_4^{2-}	CO_3^{2-}	HCO_3^-	Total salinity	Water type
	mg/L	mg/L	mg/L	mg/L	mg/L	mg/L	mg/L	mg/L	g/L	
Chang-4 + 5	39439	8849	1266	1693	80907	0	0	173	132.33	$CaCl_2$
Chang-6	6985	19002	18	120	44380	0	0	200	70.70	$CaCl_2$
Chang-8	16914	6712	347	672	39228	0	24	461	64.36	$CaCl_2$

TABLE 1-4 Composition of Associated Gases

Reservoir	CH_4	C_2H_6	C_3H_8	iC_4H_{10}	nC_4H_{10}	iC_5H_{12}	nC_5H_{12}	iC_6H_{14}	nC_6H_{14}	CO_2	N_2	Air	Hydrocarbons
Chang-4 + 5	71.27	10.92	10.45	1.11	2.29	0.32	0.52	0.16	0.17	0.50	2.29	0.33	97.21
Chang-6	57.16	13.25	16.39	1.92	5.89	0.87	1.61	0.42	0.52	0.10	1.86	1.34	98.04
Chang-8	53.15	13.42	20.28	3.01	6.51	0.80	1.19	0.33	0.16	0.12	0.96	5.98	98.85

account for 54.3%, Type-C reservoirs account for 8.6%, and Type-D reservoirs account for 3.4% (Table 1-5). On the whole, the ultralow-permeability reservoirs in Changqing have a high movable fluid saturation, with an average of 45.2%, mostly of Type A and B, which account for 87.9%, thus predicting great development potential.

As to the oil-bearing layers, the Chang-8 reservoir has a higher movable fluid saturation, up to 51.4%, while Chang-4 and Chang-5 have relatively lower saturations (Table 1-6).

1.1.3.8 Features of Waterflood Efficiency

Our calculation with the data acquired through lab imbibition shows a nondimensional net water input of 0.29% ~ 5.42%, which demonstrates that the reservoir is weakly water-wet to neutral, and our lab test of water displacing oil shows that the displacement efficiency at the water-free stage is 20% ~ 28.5%, with an ultimate efficiency of 47.7% ~ 59.5% (Table 1-7).

TABLE 1-5 Classification of Movable Fluid Saturations in Ultralow-Permeability Reservoirs of Changqing

Type	Classification criteria (%)	Number (core samples)	Percentage (%)	Porosity (%)	Permeability $(10^{-3} \mu m^2)$	Saturation of movable fluids (%)
I	>50	39	33.6	12.8	6.32	58.5
II	30 ~ 50	63	54.3	11.8	0.70	41.9
III	20 ~ 30	10	8.6	10.2	0.19	25.9
IV	<20	4	3.4	7.9	0.05	14.6
Total		116		11.9	2.53	45.2

TABLE 1-6 Correlation of Movable Fluid Saturations in Different Formations

Formation	Number (core samples)	Porosity (%)	Permeability $(10^{-3} \mu m^2)$	Saturation of movable fluids (%)
Chang-4 + 5	19	10.9	0.56	42.4
Chang-6	58	12.4	0.96	40.8
Chang-8	39	10.5	1.56	51.4

TABLE 1-7 Statistics of Water Displacing Oil Test

Reservoir	Number (core samples)	Irreducible saturation (%)	Irreducible oil saturation (%)	Two-phase flow saturation (%)	Displacement efficiency (%)			
					Water-free stage	Stage with 95% water	Stage with 98% water	Ultimate
Chang-4 + 5	11	32.9	35.3	31.8	21.8	34.7	38.32	46.2
Chang-6	153	35.1	29.3	33.6	25.5	44.9	50.2	59.5
Chang-8	38	32.5	35.6	31.9	17.9	31.9	36.5	44.9

1.2 FEATURES OF CONVENTIONAL WATERFLOODED DEVELOPMENT

1.2.1 Water Absorptivity of Reservoirs

The fluids in an ultralow-permeability reservoir will not flow unless the driving pressure difference reaches a certain critical value (i.e., the startup pressure difference).

1.2.1.1 Waterflood Pressure

The ultralow-permeability reservoirs in Changqing contain mostly acid-sensitive minerals but few water-sensitive minerals, showing high critical flow velocities, neutral to weakly water-wet wettability, and relatively low viscosities of formation crude. Most of these reservoirs are sandstones of mixed or inverted rhythms, which means these reservoirs provide good conditions for waterflooding.

Since fractures often have higher permeabilities, the ultralow-permeability reservoirs where microfractures are well developed usually have strong water absorptivity and require relatively low flooding pressure, which remains comparatively stable in the flooding process. Take the Chang-6 reservoir in Ansai, for example. The flooding pressure at the beginning was about 6 MPa, and by the end of 2006 it reached 7 MPa. Such stability of flooding pressure is good for advanced waterflooding (Fig. 1-16).

1.2.1.2 Index Curve Properties of the Injection Wells

Fig. 1-17 shows the apparent injectivity curves of waterflooded blocks in the Triassic reservoirs, where the indexes remain quite stable (3 m^3/d MPa $\sim$ 4 m^3/

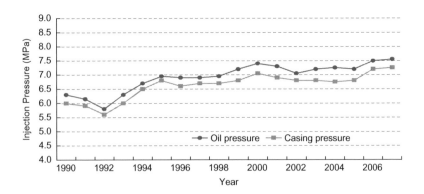

FIGURE 1-16 Waterflooding pressures profile in the Ansai Oilfield.

d MPa), but because natural microfractures are well developed in these formations, the curves will reach a breakpoint once the injection pressure exceeds the starting pressure of the microfractures. Take Well Wang-17-7, for instance. The breakpoint pressure is 8.0 MPa, and the water injectivity index is 4.0 m³/d MPa before it, while reaching 16 m³/d MPa after it (Fig. 1-18). Therefore, the injection intensity should be reasonable so as to avoid fracture opening and water breakthrough due to excessive injection pressure.

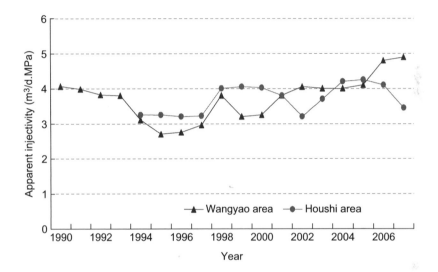

FIGURE 1-17   Apparent water injectivity index of Wangyao.

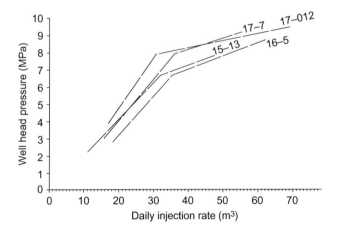

FIGURE 1-18   Water injectivity index curve in Ansai and Houshi blocks in Ansai.

1.2.1.3 The Entry Profile of the Injection Wells

Nonfracturing Injection Increases the Sweep Volume of Injected Water

At the initial stage of development in the Ansai Oilfield, techniques of fracturing, unloading and flushing were used before water injection. In 1990, the oilfield began to experiment with nonfracturing injection in 44 wells, with an average well-head pressure of 6.1 MPa, merely 0.2 MPa higher than in the 23 neighboring wells using fracturing injection. The result is that nonfracturing injection produced a higher water injectivity than fracturing injection. It can be concluded, then, that nonfracturing injection can not only improve the sweep rate, but also lower the costs, resulting in a better development effect.

Natural Fractures Intensify the Areal Conflict

As natural microfractures are well developed in the Triassic reservoirs, some blocks, after fracturing and water injection, may experience water breakthroughs along the fractures when the injection pressure exceeds the fracture-opening pressure, causing areal conflicts and vertical disproportion along the injector-producer profile.

In such zones, the water injectivity curve may reach a breakpoint, at which the water injectivity index may rise abruptly; or it may be a steady straight line, indicating a large water injectivity index. Some entry profiles show the injectivity curve like spike-teeth (leptokurtic curve) (Fig. 1-19). Such changes in the curves imply a dual effect of water injection on oil development. On the one hand, the producing wells on the fractures may

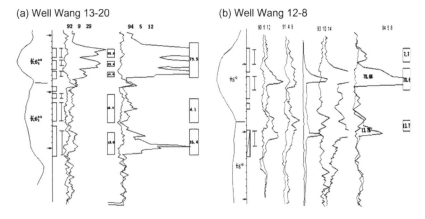

FIGURE 1-19   Changes in water injection IPR curves in the Ansai Oilfield.

have a quick response and water breakthrough, with the waterline advancing at a rate of 0.43 m/day ~ 4.35 m/day. Some specific wells may even experience explosive flooding. But on the other hand, the wells on the lateral sides of the fractures may have a slow response, or even no response for a long time. This dual effect intensifies areal conflicts in waterflood development. At the same time, the water drive in the layers with few fractures developed may have poor results for oil production.

1.2.2 Features of Oil Producer Responses

1.2.2.1 Features of Responses

In the waterflood development of ultralow-permeability reservoirs in Changqing, the oil producers often have a response time of three to six months. The response rates of Chang-8, Chang-6, and Chang-4 + 5 range between 54.3% and 86.2% after one year of injection, but the areal responses are different, with some wells responding very slowly. Take the Wangyao block of Ansai, for example. The response rates of wells in its middle and western parts reached up to 86%, with the productivity rising about 2 t/d, but those of the wells in the eastern part remained only 43%, with the productivity rising only less than 0.5 t/d. Some oil wells in this part are still in a state of low pressure and low output.

The waterflood response in these reservoirs can be described as "four indexes rise, one remains stable, and one declines." That is to say, with water injection, the reservoir pressure, the per-well daily output, the working fluid level, and the pump efficiency tend to rise, the water-cut keeps stable, and the oil/gas ratio declines. For example, after water injection became effective in the Wangyao block, the average per-well daily output rose to 4.03 t/d, the water-cut remained at 16%, and the working fluid level was 651 m, with an average producing rate of 1.89% and a recovery factor of 7.61%. Compared with that from the wells with no response to water injection, the per-well output of these wells increased by 1.6 times (the output of a non-response producer is 2.5 t/d, at a withdrawal rate of 0.78%). In addition, most of the response producers have a longer period of steady output (usually 2 ~ 3 years).

As the flood becomes effective, some producers begin to suffer water breakthroughs, falling into three types, i.e., the porosity type, the pore-fracture type, and the fracture type. Different types of water breakthroughs may have different features. Take Wangyao, for example. In the wells experiencing the porosity-type water breakthroughs, the injected water that flows through pores takes a longer time (684 days) to produce an effect, and advances slowly (0.61 m/day). As a result, with the water-cut rising at a low speed (6.09%), these producers can keep a relatively high productivity after water breakthroughs. The wells with the fracture-type breakthroughs have a shorter breakthrough response time (97 days), with a higher velocity of waterline advancement (4.06 days) and water-cut rise (47%), and a sharper decline of productivity. A few wells may even experience explosive flooding. The indexes in the wells with the pore-fracture-type breakthroughs lie in between those of the former two types, with the rise of the water-cut influenced by the pressure and flow rate of water injection.

Differences in water breakthroughs in oil wells are related to patterns of oil and water movement, which is controlled by geologic features of reservoirs. In other words, different sedimentary facies may lead to different patterns of oil and water movement.

Breakthroughs in the Facies of the Underwater Distributary Channel

The injected water rapidly tongues forward along the main zone, with an average breakthrough time of 354 days, and an advancing velocity of waterline of 1.83 m/day, twice as high as in other facies. In addition, the water-cut rises at a rate of 21.4%.

Breakthroughs in the Mixed Facies of the Distributary Channel and River-Mouth Bar

Along with the forward aggradation and traction of the delta, the distributary channel aggrades above the river-mouth bar, forming a straight, thin, and narrow sandbody. In such facies, the injected water may first tongue forward along the distributary channel, but obviously at a lower speed. The water breakthrough time is 634 days, with the advancing at a rate of 0.65 m/day and the water-cut rising at a rate of 8.16%. After the water injection becomes effective, the average output reaches 3.62 t/d, and the producing rate reaches 1.8%, thus showing a better development efficiency.

Breakthroughs in the Facies of the River-Mouth Bar

Not only are sandbodies consisting of river-mouth bars usually widely distributed, stable, and thick, they also have good physical properties and allow balanced waterflooding. As a result, water injection in these sandbodies can produce a prompt response in the producing wells, with the water-cut rising slowly. After the water injection becomes effective, the average daily output reaches 4.3 t/d, the producing rate reaches 2.03%, and the recovery factor reaches 8.07%.

1.2.2.2 Factors That Influence Well Responses

Waterflood effects in ultralow-permeability reservoirs are different from those in high-permeability ones. Predicted from the relations between the radius and the time of pressure-wave effects, the breakthrough time in ultralow-permeability reservoirs is $3 \sim 6$ months. But according to the dynamic data acquired from actual reservoir development, it takes $4 \sim 8$ months on average for waterflood to become effective in such reservoirs. And the wells where the reservoirs have poor physical properties may even have no response for a long time.

The Influence of Fractures

The differences between the permeabilities of the matrix and those of the fractures may lead to severe areal conflicts in the process of waterflooding. When the injectors and producers are both located in the direction of fracture extension, the producer may suffer from water breakthrough too soon. Sometimes the directional advancement of water will cause disastrous consequences, resulting in explosive flooding in producers. Wells in the direction perpendicular to the fractures, however, may hardly receive any injected water because of the very low matrix permeabilities. Such ineffectiveness of water injection may lead to low pressure and low output of the wells on the sides of fractures.

Figure 1-20 shows our experimental injection work in Well Sai-6−71 in the Ansai Oilfield, in which the producing formations are the Chang-6 reservoirs in the Yanchang Triassic Group. With permeabilities between $0.56 \times 10^{-3} \, \mu m^2$ and $4.85 \times 10^{-3} \, \mu m^2$ only, these reservoirs contain well-developed microfractures, generally in the NE $\sim$ SW directions. Water injection for these wells began in June 1987, at an injection rate of $30 \, m^3$/day, which was kept stable. From July 1989 to December 1991, the rate was raised to over $40 \, m^3$/day, but no more than $60 \, m^3$/day. Half a year after the injection rate was raised, Well Sai-6−9, which lies the furthest (450 m) from Well Sai-6−71, showed water breakthroughs. At the same time, Well Sai-6−6, which is located in the lateral direction of fractures and is the nearest (146 m) to well Sai-6−71, showed a very good effect from water injection. The daily output from the well rose from only 2.39 tons before waterflooding to 4.25 tons after water injection became effective, an increase of 78%, and remained steadily over 3.0 tons for more than 10 years. In December 2005,

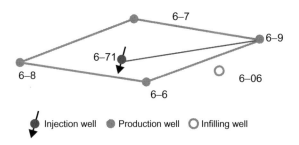

FIGURE 1-20   Well pattern of Well Sai 6−71.

the daily output from the well was 2.0 tons, with a water-cut of 48.3%. By that time, the well had produced a total of 1.83×10^4 tons of oil.

The Influence of Well Spacing

In the development of reservoirs with well-developed fractures where inverted nine-spot diamond and rectangular patterns are adopted, well spacing is also an important factor influencing the effectiveness of water injection on oil producers. In other words, the larger the well spacing, the longer time it takes for water injection to become effective. In some zones where fractures are well-developed, the large-spaced oil wells may even have no response to injection for a long time because the water injected often channels off along the fractures.

Take the Pingqiao block of the Ansai Oilfield, for example. An inverted nine-spot injection pattern on a square of 250×250 m was used in 1988. Due to the developed microfractures, water injection had already produced an effect in the producers along the lines of 17 fractures by the end of 2006. However, the producers on the sides of the fractures did not respond for a long time. Although kept stable through enhanced water injection, the per-well output remained at a low level, only 1.5 t/d. In 1999, an inspection well, Ping-1, was drilled 80 m away from the fracture on its side, but no water was seen in the coring. After Ping-1 was put into production, the initial daily output was 4.58 tons, with a water-cut of 10.2%. Then two infill wells, i.e., Ping 30−391 and Ping 30−392, were drilled, which formed a small-spaced test well group, with a spacing of 50~100 m. The infill wells maintain a stable production of 2.2 t/d, with a comprehensive water-cut of 8%.

The Influence of Injection-Production Parameters

To solve the problems of the injected water channeling along the fractures, the water-cut rising rapidly or even explosively in the producers in the main directions, and the producers on the sides being low pressure, low production, and low recovery for a long time, an experiment of enhanced injection well rows was carried out in recent years that has improved the development efficiency to a certain degree in the fracture-developed areas.

The enhanced water injection along fractures, which was carried out in the east of the Pingqiao and Wangyao blocks, showed a positive result in that it improved the flooding seepage in the producers on the lateral sides of waterlines and increased the per-well productivity. On the one hand, the pressure in the producers on the lateral sides went up. For example, the pressure in the producers on the lateral sides of fractures in Pingqiao rose from 4.6 MPa in 1988 to 6.92 MPa in 2005 after water was injected along the fractures. On the other hand, the outputs of producers on the lateral sides of fractures increased and the output decline slowed down. The average daily outputs of the 21 producers on the lateral sides of fractures in Wangyao rose from 1.15 tons in January 1998 to 2.19 tons in October that year. In 10 of the 21 producers, the average daily outputs rose from 1.3 tons before injection to 2.19 tons after the injection became effective. In addition, the output decline of producers on the lateral sides of fractures in Pingqiao obviously slowed down.

The Influence of Interlayer Connectivity

Apart from the three factors discussed, the oil-layer connectivity also has an influence on the effectiveness of water injection and how soon it produces a response in the oil wells. For example, among the four injection blocks in the Ansai Oilfield, while Pingqiao and Wangyao contain only single oil layers, the central and southern parts of Xinghe and Houshi contain multiple oil layers. The development method of opening multiple layers through one layer by using commingled water injection for all wells created the problem of poor or no water absorption into the layers in the vertical direction, resulting in little or no effect on production wells. For example, in the Xinghe block where capacity building was initiated in 2001, some producers still had no obvious response after five years of water injection, with a per-well output of only 1.0 t/d.

1.2.3 Features of Producer Deliverability

1.2.3.1 The Initial Deliverability

The producers in the Triassic reservoirs show little or no natural deliverability because of their ultralow permeabilities and low initial pressures. The drill-stem testing in Ansai with oil-based mud and foam underbalanced drilling test indicated that the initial output of a producer was just $0.3 \sim 0.5$ t/d.

Therefore, no commercial oil flow could be obtained without fracturing treatment of these reservoirs. The fracturing testing with a small-scale gravel input of $5\,m^3 \sim 6\,m^3$ in the early 1970s resulted in an initial output of $7 \sim 8$ t/d. With 30 years of development, the fracturing techniques and strategies have become more mature, with a gravel input per well reaching $30\,m^3 \sim 40\,m^3$, a test output of $14 \sim 20$ t/d and an initial output over 4 t/d.

1.2.3.2 Patterns of Deliverability Changes

Both the Reservoir Pressures and the Outputs Decline Quickly during Initial Production

The conventional "depletion" method of development through elastic and dissolved-gas drive in the ultralow-permeability reservoirs, which, dominated by lithologic control, have insufficient natural energy, caused a severely insufficient supply of fluids and a rapid decline of the low deliverability. For example, when the reservoir pressure of well Sai-6 in Ansai declined from 9.1 MPa to 6.3 MPa, the recovery was just 0.71%, and it further declined to 3.94 MPa when 1% of geological reserves were recovered. The per-well output of the 22 producers in the pilot development zone of Ansai, which were put into production in March 1989 without water injection, declined from 3.2 t/d to 2.58 t/d, at an annual rate of 25.8%, and further to 1.75 t/d at the end of 1990, at an annual rate of 32.2%. At the same time, the per-well output of the 55 wells in the industrial development zone declined from 4.32 t/d to 2.85 t/d only one year after commissioning, with an annual rate of 32.6%.

Producers at Different Stages of Water-Cut Show Different Decline Rules

An analysis of the decline history over the years of waterflood development of Triassic reservoirs such as in the blocks of Wangyao, Pingqiao, Houshi, and Xinghe in the Ansai Oilfield, those of Wuliwan, Panguliang, and Dalugou-2 in the Jing'an Oilfield, and the 152-block in the Huachi Oilfield shows that the output decline may have the following features.

The output tends to increase at the stage of low water-cut. An oilfield with a water-cut lower than 20% is at the stage of waterflood response when the output keeps stable or increases, producing a good development effect. For instance, the old wells brought into production before 1999 in Xinghe and those brought into production before 2000 in Wuliwan, with a water-cut of 17.5% and 14.3%, respectively, at the end of 2006, continued to have a stable or increasing output with effective waterflood (Figs. 1-21 and 1-22).

The output begins to decline at the stage of intermediate water-cut. A water-cut of 20% means an oilfield enters the stage of intermediate

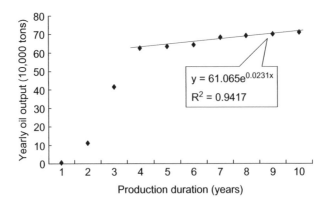

FIGURE 1-21 The decline curve of old wells brought into production before 1999 in the Xinghe block.

water-cut, when the output begins to decline, usually at a high rate. In the Wangyao block, the output began to decline when the water-cut reached 20%. When it further rose to 53.8% at the end of 2006, the output declined exponentially, at a rate of 10.9%. In the Hua-152 block, the output began to decline when the water-cut reached 34%. When it furthered rose to 46.3% at the end of 2006, the output declined exponentially, at a rate of 10.6% (Figs. 1-23 and 1-24).

1.2.4 Changes in Productivity Indexes (PIs)

The neutral and weak water wettability of the ultralow-permeability reservoirs in the Ordos Basin, accompanied by water sensitivity, water lock and

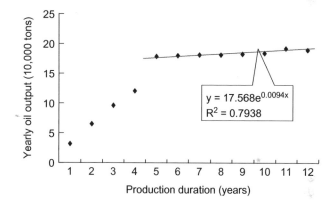

FIGURE 1-22   The decline curve of old wells brought into production before 2000 in the Wuliwan block.

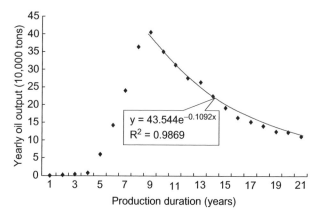

FIGURE 1-23   The decline curve of old wells brought into production before 1999 in the Hua-152 block.

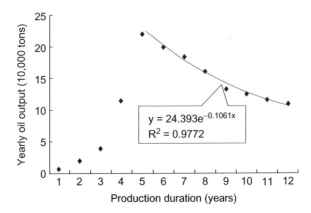

FIGURE 1-24 The decline curve of old wells brought into production before 1995 in the Wangyao block.

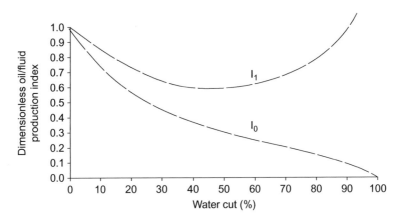

FIGURE 1-25 Dimensionless oil/fluid production index curves in Ansai.

speed-sensitivity in the waterflooding process in some areas, the delayed injection and the dropped reservoir pressures, cause an irreversible decline of permeabilities. Therefore, the oil-water relative permeability curves show an increase of the water saturation, an abrupt decline of oil-phase permeability and a gradual increase of water-phase permeability, which is at a maximum of 0.6. At last, the PI declines while the water-cut rises (Figs. 1-25 and 1-26). Actual data from the Wangyao block in Ansai indicate that, by the end of 2005, after a long period of production, the rate of reserve recovery reached 13.7%, the composite water-cut rose from 29.0% to 49.8%, the fluid productivity factor declined from 0.88 m^3/(day.MPa) to 0.47 m^3/(day.MPa) and the oil productivity factor declined from 0.69 t/d MPa to 0.24 t/d MPa.

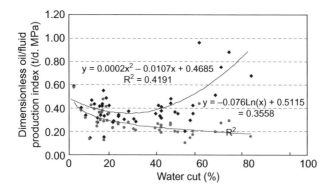

FIGURE 1-26   Dimensionless oil/fluid production index curves in Wangyao.

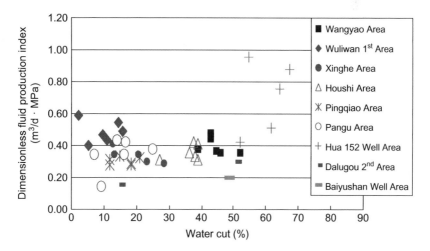

FIGURE 1-27   Fluid PI curves in the waterflood production of Triassic reservoirs.

An analysis of the PI curves in the nine flooded blocks from the three oil-fields of Ansai, Jing'an, and Huachi shows that different blocks at the same level of water-cut have PIs in the same range, but blocks at different stages of water-cut have PIs in different ranges. In addition, a block with a higher water-cut has low PIs, which is consistent with the PI variations in ultralow-permeability reservoirs. Further analysis also indicates that the relations between the productivity indexes and the water-cuts in the different blocks at different stages are similar to each other (Figs. 1-27 and 1-28).

So it can be concluded that in the waterflood production of ultralow-permeability reservoirs in the Ordos Basin, the PI changes follow the same trajectory, whose main feature is that productivity indexes, usually low, will decline as the water-cut increases, and when the water-cut reaches about 40%, the fluid PIs will increase but the oil PIs will continue to decline.

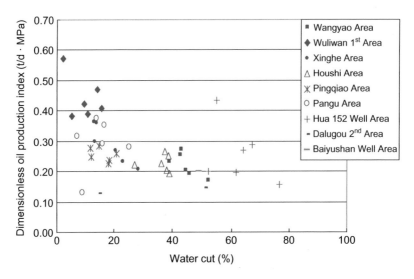

FIGURE 1-28 Oil PI curves in the waterflood production of Triassic reservoirs.

1.3 THE INTRODUCTION OF ADVANCED WATER INJECTION TECHNOLOGY

Practices show that after the ultralow-permeability reservoirs are put into production, the reservoir pressure may drop quickly without timely energy addition. As a result, the output will decline rapidly, the PIs will fall at an annual rate between 25% and 45%, and withdrawing 1% of geological reserves will cause a drop of reservoir pressure by $3 \sim 4$ MPa. What is worse is that even if the reservoir pressure is raised later, the production and productivity index will be difficult to restore. This is caused by the pressure sensitive effect, i.e., the fluid-solid coupling effect.

The obvious elastoplasticity of the ultralow-permeability reservoirs leads to high-pressure sensitivity of the formations. In other words, the reservoir porosity and permeability will decline rapidly when the pore pressure drops but will be difficult to restore when the pore pressure rises again. For example, the permeability can decline by $70\% \sim 80\%$ but can be restored by less than $20\% \sim 30\%$. This is the main reason why the producer outputs and the PIs of the ultralow-permeability reservoirs can rarely be restored after declining. From both the production practice and theoretical studies angles, therefore, it is necessary to avoid the drop of reservoir pressure in order to maintain the initial productivity and to achieve good development efficiency in ultralow-permeability reservoirs. To do this, the development mode of advanced injection should be adopted.

Scientists and technologists from the Changqing Oilfield, while addressing the reservoir features of big kick-off pressure differences, high-pressure sensitivities, low-pressure coefficients, and slight differences between reservoir pressure and saturation pressure, creatively proposed a theory of advanced water injection through in-depth studies of flow patterns and constant technological practices. At the same time, they also investigated the technological policies of injector-producer distance, waterflood timing, reservoir pressure to be maintained, injection pressure, and water intake per unit thickness, which formed supporting technologies for advanced waterflood production in ultralow-permeability reservoirs.

Advanced injection is an injection-production method in which the injectors start water injection before the producers begin to operate, and the reservoir pressure should be kept at a certain level to establish an effective system of displacing pressure.

1.3.1 Functions of Advanced Water Injection

Advanced waterflood development of the ultralow-permeability reservoirs in the Changqing Oilfield reasonably provides additional energy for the reservoirs, increases the reservoir pressure, and reduces damages to permeability caused by pressure drop. Thus, a large production pressure difference can be created for the producer to offset the threshold gradient after it begins operation, keeping productivity high for a long time. In addition, advanced injection can prevent the physical properties of the crude from getting worse, effectively facilitate the smooth flow of the oil in porous media, enlarge the sweep volume of injected water and ultimately, increase the oil recovery.

1.3.1.1 Maintaining High Reservoir Pressure to Build an Effective Displacing Pressure System

Ultralow-permeability oilfields have threshold pressure and nonlinear fluid flows. Experiments both at home and abroad have shown that, when reservoir permeabilities decline to a certain level, the flow characteristics no longer follow Darcy's Law, and there exists a certain threshold pressure gradient. As a result, to build an effective displacing pressure system, the displacement gradient must exceed the threshold gradient between the producer and the injector.

In ultralow-permeability reservoirs, the oil layers will not begin to produce oil unless the displacing pressure gradient becomes higher than the threshold gradient. On account of the threshold gradient, the displacing pressure difference may cause different effective driving distances in different layers with varying permeabilities in a certain reservoir, which can be expressed as

$$\frac{p_{inj} - p_w}{\ln \frac{R}{r_w}} \cdot \frac{2}{R} = aK^{-b} = 0.0608K^{-1.1522} \tag{1.1}$$

For any layer, when the effective driving distance is shorter than or equal to the injector-producer distance, the oil cannot be displaced effectively. We can obtain the minimum effective permeability from Eq. (1.1):

$$K = \left(\frac{p_{inj} - p_w}{\ln \frac{R}{r_w}} \cdot \frac{2}{aR} \right)^{-\frac{1}{b}} \quad (1.2)$$

NIU Yanliang and LI Li from the Daqing Oilfield conclude that in a specific reservoir with a reasonable development mechanism, the relations between the effective displacement and the well-spacing density can be expressed as

$$E_M = 1 - m(K - K_0)^d \quad (1.3)$$

Putting the minimum reservoir permeability K into Eq. (1.2) results in an equation for the rate of waterflood displacement:

$$E_M = 1 - m(K - K_0)^d = 1 - m \left[\exp \left(\frac{p_{inj} - p_w}{\ln \frac{R}{r_w}} \cdot \frac{2}{aR} \right)^{-\frac{1}{b}} - K_0 \right]^d \quad (1.4)$$

where
p_{inj}—injection pressure, MPa
p_w—producer bottom pressure, MPa
R—maximum drainage radius of producer, m
r_w—wellbore radius, m
K—reservoir permeability, $10^{-3} \, \mu m^2$
K_0—minimum permeability of effective thickness obtained through gas logging, $10^{-3} \, \mu m^2$
a, b, m, d—relevant coefficients

Figure 1-29 shows that the water-displaced oil is proportional to the displacing pressure gradient. In other words, the higher the displacement gradient, the more oil the injected water can displace. In addition, a shorter injector-producer distance may help produce more oil displaced by water.

1.3.1.2 Reducing Damage to Permeability Caused by Pressure Drop

Oil wells may drain the fluids from the formations around the wellbore. Since the ultralow-permeability reservoirs have insufficient natural energy and poor energy conductivity, the energy consumed by wellbore drainage is very hard to restore in a short time. As a result, the reservoir pressure may soon fall, thus deforming the reservoir matrix and lowering the

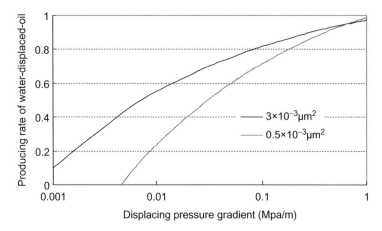

FIGURE 1-29   Relations between the water-displaced oil and the pressure displacement gradient.

permeabilities, which in turn speeds up the pressure drop and output decline. Even if the reservoir pressure can be restored to its original level through water injection, the reservoir permeability cannot. Our experiments with elastic-plastic deformation of rocks show that the physical properties of the ultralow-permeability reservoir will change as the reservoir pressure changes. Significant pressure changes may transfer the rock deformation from elastic to plastic, which may also worsen the physical properties of the reservoirs.

To sum up, in the development of ultralow-permeability reservoirs, advanced injection can help maintain the reservoir pressure, thus minimizing the effect of pressure sensitivity of the reservoir media, of the elastic-plastic deformation of rocks, and of the threshold pressure gradient. In conclusion, from the viewpoint of reservoir pressure maintenance, advanced injection is a necessity for the development of ultralow-permeability reservoirs.

1.3.1.3 Preventing Changes in the Physical Properties of the Oil in Place

On account of poor physical properties, high filtrational resistance, and big pressure losses in the ultralow-permeability reservoir, the pressure will drop down below the bubble point a short time after commissioning, and the oil in place will begin outgassing. While some dissolved gases flow to the well bottom, others remain in the formation. The outgassed crude then will build up a new phase balance under new reservoir pressure. Figure 1-30 shows that the pressure decline lowers the content of dissolved gases in the oil on the one hand, but on the other hand, increases the oil density and viscosity, thus making the remaining oil less movable and the development more difficult.

Conventional injection can restore the dropped reservoir pressure so that the oil in place will acquire new physical properties in the new balance. Even if under the same pressure, however, the crude density and viscosity will rise while the volume factor will reduce because of earlier degasification. As a result, the filtrational resistance will increase, and the crude output will fall under the same producing pressure difference.

Advanced injection, on the other hand, can prevent the reservoir pressure from dropping at the very beginning and maintain it around the initial level

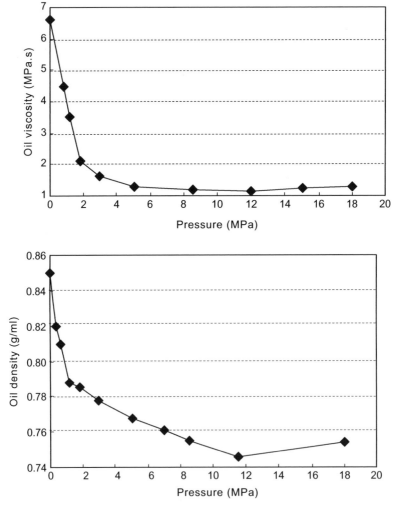

FIGURE 1-30   Relations between the reservoir pressure and the oil density and viscosity in the Xifeng Oilfield.

so long as the injection and the production can be balanced. This can prevent the oil properties from getting worse, thus helping to increasing the per-well output.

1.3.1.4 Preventing the Flow Matrix from Being Plugged

When the reservoir pressure drops down below the bubble point in oil production, the oil in place begins to outgas. Since the radius of some pore throats is very short in ultralow-permeability reservoirs, it is hard for the escaped gasses to pass through these tiny pores. As a consequence, they have to reside within the formation, forming an air lock or Jamin's effect, which reduces the effective flow paths. In this situation, the oil output will decline even if the producing pressure is kept unchanged. Advanced injection, on the other hand, can maintain the reservoir pressure in the neighborhood of its initial level, thus preventing the oil in place from outgassing and leading to a higher per-well output.

1.3.1.5 Improving the Oil Permeability

Figure 1-31 presents curves of relative oil/water permeabilities under different displacing pressures in the ultralow-permeability cores. It can be seen from the figure that when the displacing pressure increases, the water permeability curve changes little while the oil permeability curve moves upward. This is because there is an obvious non-Darcy effect in the two-phase fluid flow zone when the permeability is very low. At the same time, in our experiments, the interstitial oil saturation (IOS) reduces a bit when the displacing pressure is raised.

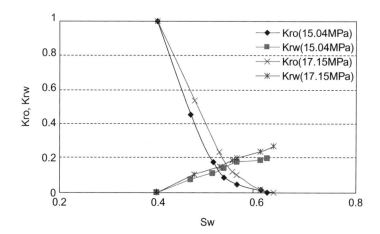

FIGURE 1-31 Oil and water relative permeability curves under different displacing pressures.

1.3.1.6 Increasing the Sweep Efficiency of the Injected Water to Enhance Oil Recovery

Before the oilfield is developed, the original reservoir pressure is in a balanced state, with the pressure at each point almost the same. For this reason, the water injected before commissioning can advance evenly outward to the formations surrounding the bottom hole. First, the injected water flows into the high-permeability zones of little filtrational resistance. When it encounters higher pressures in these zones, the resistance preventing water from flowing increases. Then the pressure difference between layers of high and low permeabilities forces the water to flow into the zones of lower permeabilities. With the pressure of the lower-permeability reservoirs thus increased and the difference between layers of high and low permeabilities reduced, the sweep efficiency of the injected water is improved, and so is the efficiency of oilfield development. Our water displacement experiments in low permeability oilfields show that an increase in displacement gradient may produce a higher displacement efficiency, thus increasing or maintaining the reservoir pressure at a high level, increasing the displacing pressure and producing pressure differences. In this way the capillary tube forces and other resistant forces can be overcome and the oil in thinner porous paths displaced.

Using conventional delayed injection to develop ultralow-permeability reservoirs, producers tend to take out the oil from the zones of higher permeabilities first. However, due to their stronger filtrational resistance, poor deliverability, and quick energy depletion, the pressure in these zones would drop promptly. When the injectors started to inject water, the water would break through into the higher-permeability zones of weaker filtrational resistance. Such breakthroughs accompanied by great pressure drop in higher-permeability zones would lead to serious water fingering, thus reducing both the sweep efficiency and the displacement. On the other hand, the method of water injection before commissioning (advanced injection) can keep the water moving evenly in the formations because the reservoir pressure is still in an original balance, with the pressure at various points almost the same. The water injected before commissioning would also break through first into the higher-permeability zones of weaker filtrational resistance. But as the pressure in these zones rises, the injected water will turn to the lower- or ultralow-permeability zones, thus effectively increasing the swept volume.

Our water-displaced oil experiments under different displacing pressures show that when the displacing pressure gradient is enhanced, oil in thinner pores can be forced out, with a rise of the displacement efficiency (Fig. 1-32). In addition, as has been discussed, advanced water injection can effectively increase the swept volume and enhance oil recovery.

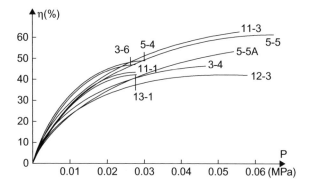

FIGURE 1-32   Curves of displacing pressure and efficiency in Ansai.

Our numerical modeling shows that advanced injection can increase the recovery efficiency by $3\% \sim 5\%$.

1.3.1.7 Slowing Down Output Decline

The key to the development of ultralow-permeability oilfields is to enhance the per-well output and extend the duration of stabilized output so as to improve their economic conditions. The basin-wide application of advanced injection in Ordos succeeded in reducing the decline of the initial output. Take the Baima block of the Xifeng Oilfield, for example. Data from its capacity-building wells drilled in 2002 and 2003 show that their total output decline by the third year was, respectively, 8.5% and 5.0%. However, the total decline of the capacity-building wells in the Triassic reservoirs not using the advanced injection reached 14.2% in the first four years.

Another case is from the advanced-flooded blocks in Wuliwan and Dalugou of the Jing'an Oilfield. Thanks to the effective pressure displacement system built by advanced injection, the initial per-well output in these blocks increased by 1 t/d $\sim$ 2 t/d and remained there for a long time (Figs. 1-33 and 1-34).

1.3.1.8 Improving Producer Deliverability

On the basis of the mechanism of the pressure-sensitive effect of deformable media, we built a mathematical model of the threshold pressure gradient and a model of the non-Darcy filtration for boundary layers in an ultralow-permeability reservoir. Our innovative advanced injection technique, aimed at raising the reservoir energy, overcoming the threshold pressure gradient, and building an effective pressure displacement system, helped raise the average per-well deliverability by $15\% \sim 20\%$.

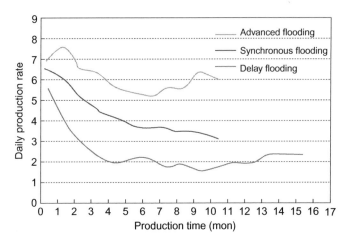

FIGURE 1-33　Diagram of advanced-flooding effect in Wuliwan.

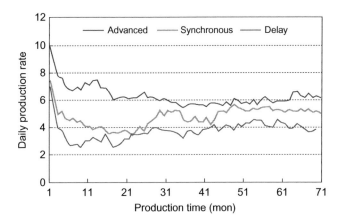

FIGURE 1-34　Diagram of advanced-flooding effect in Dalugou.

1.3.1.9 Greatly Enhancing the Maximum Injector-Producer Distance and the Effective Coverage

When the reservoir pressure is higher than the threshold pressure for a microfracture to open, the reservoir permeability will increase greatly. Both theoretical and case studies indicate that the threshold pressure is negatively proportional to the permeability. In addition, in the development of ultralow-permeability reservoirs, the maximum injector-producer distance is positively proportional to the driving pressure (Fig. 1-35) but negatively proportional to the threshold pressure gradient.

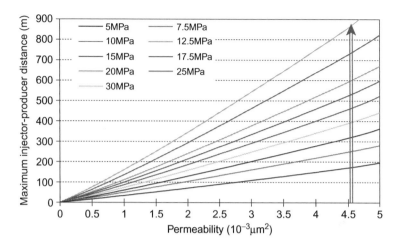

FIGURE 1-35   The relationship between permeability and maximum injector-producer distance with different pressures.

1.3.2 Applicability of Advanced Water Injection

Advanced injection can be applied to the development of reservoirs under the following conditions:

Low original pressure coefficient in the reservoir

The edge water and bottom water in the Yanchang Group of Triassic reservoirs are inactive. The lack of natural energy leads to a quick drop of output, pressure, and the producing fluid level at the initial stage of production. Take Chang-6 in Dalugou block 2 as an example. With a buried depth averaging about 1680 m, the reservoir has an original pressure of 10.51 MPa, a pressure gradient of about 0.63 MPa/100 m and a saturation pressure of 8.8 MPa. Another example is the Chang-8 reservoir in Xifeng, with a buried depth of 2000 m ~ 2100 m, an original pressure of 18.1 MPa, a pressure gradient of 0.883 MPa/100 m, and a saturation pressure of 11.6 MPa. The Chang-4 + 5 reservoir in Jing'an has an original pressure of 10.48 MPa, a depth 1715 m in its middle, and a pressure gradient of 0.658 MPa/100 m. And the same reservoir in the west block of Nanliang has an original pressure of 12.71 MPa, a depth of 1851 m, a pressure gradient of about 0.687 MPa/100 m, and a saturation pressure of 11.56 MPa. All these reservoirs are low pressure and have ultralow permeability, and reservoirs in the blocks of Wuliwan and Panguliang have similar features.

Fine pore throats

The 19 cores from the Chang-4 + 5 reservoir have an average pore throat value of 11.32Φ, with a median throat radius of 0.21 μm and a median pressure of 4.98 MPa. The 10 cores from the Chang-6 reservoir have an average pore throat value of 11.5Φ, with a median throat radius of 0.17 μm and a median pressure of 5.47 MPa. The 90 cores from the Chang-8 reservoir have an average pore throat value of 11.54Φ, with a median throat radius of 0.286 μm and a median pressure of 4.43 MPa. Observe that the median throat radius of these reservoirs is usually shorter than 0.5 μm, showing a high median pressure and strong filtrational resistance. The adoption of advanced injection, therefore, can help maintain the reservoir pressure and shorten the response time.

High threshold pressure gradient

One major problem in the Changqing Oilfield is a lack of producing energy. Quite a few reservoirs have a pressure coefficient of only $0.6 \sim 0.7$, which turns out to be the direct kinetic cause of low output per well. Advanced injection aims at solving this problem to improve the per-well output.

The general mentality for advanced injection is to pour water into the injectors before the producers are put into production. At this lead-in stage, the method of "flooding but not producing" can effectively raise the reservoir pressure. When the pressure in the reservoir reaches a threshold, a reasonable injector-producer spacing can make the working gradient at any point higher than the threshold gradient, thus building an efficient displacing pressure system. The Triassic reservoirs of the Yanchang Group in Changqing have a high threshold gradient. Our experiments with 10 cores from these reservoirs show a threshold pressure gradient between 0.071 MPa/m and 0.933 MPa/m, averaging 0.678 MPa/m (Table 1-8). Under such a high-pressure gradient, it is hard to build up an efficient displacing pressure system with an injector-producer distance of 300 m in the conventional method, but advanced injection can improve the efficiency of both injection and production.

High Saturation Pressure of Oil in Place but Small Difference between Reservoir and Saturation Pressures

With the small difference between the reservoir and saturation pressures, the former can easily drop below the latter. If the flowing pressure is too much lower than the saturation pressure, the outgassing radius will increase, which will worsen the flowing conditions in both reservoirs and wellbores. Therefore, the reservoir pressure should be controlled within a reasonable range to avoid its negative influence on production.

The difference between the reservoir pressure and the saturation pressure in the Yanchang Triassic reservoirs of the Changqing Oilfield is usually small. Take for example the reservoirs of Chang-8 in mid-Baima, Chang-6 in block 2 of Dalugou and Chang-4 + 5 in Baiyushan. Their bubble-point pressures are 11.6 MPa, 8.8 MPa, and 8.5 MPa respectively, with a difference of 6.5 MPa, 1.71 MPa, and 4.21 MPa, from their respective reservoir

TABLE 1-8 Results from Threshold Pressure Measurement of 10 Cores from the Yanchang Group

Displacing velocity (ml/s)	Displacing pressure gradient at different displacing velocities (MPa/m)									
	1	2	3	4	5	6	7	8	9	10
0.03	0.539	1.530	19.14	11.81	6.440	3.524	6.703	0.511	33.86	11.32
0.02	0.389	1.052	13.73	8.568	4.478	2.405	4.842	0.346	26.84	8.064
0.015	0.305	0.820	10.92	6.753	3.451	1.877	3.848	0.267	23.30	6.392
0.01	0.216	0.556	7.683	5.029	2.389	1.286	2.848	0.181	19.50	4.855
0.005	0.114	0.288	4.319	2.840	1.223	0.687	1.599	0.095	12.87	2.778
0.0025	0.061	0.151	2.659	1.752	0.692	0.379	0.974	0.049	8.844	1.726
0	0	0	0.993	0.499	0.167	0.071	0.326	0	4.202	0.585
Pseudo-threshold gradient (MPa/m)	0.0061	0.0082	0.2217	0,1680	0.0401	0.0185	0.0951	0.0018	1.247	0.1575
Real threshold gradient (MPa/m)	0	0	0.0933	0.0499	0.0167	0.0071	0.0326	0	0.4202	0.0585

pressure. Therefore, these reservoirs are fit for advanced injection, which can help maintain the reservoir pressure.

Few Water-Sensitive Minerals

The ultralow-permeability layers in the Yanchang Triassic Group are dominated by acid-sensitive minerals mixed with a few water-sensitive minerals. Most layers show a weak or medium-to-weak water sensitivity (Chang-8 in mid-Baima), no water sensitivity (Chang-6 in block 2 of Dalugou) or weak water sensitivity (Chang-6 in Wuliwan and Chang-4 + 5 in Baiyushan) and therefore are all fit for advanced injection to achieve a better development effect.

Chapter 2

Nonlinear Percolation Theory for Ultralow-Permeability Reservoirs

Advanced Water Injection for Low Permeability Reservoirs.

Flow patterns of fluids in an ultralow-permeability reservoir are dominated by four factors: first, the composition and physicochemical properties of the fluids; second, the pore structures and physicochemical properties of the porous media; third, the flow dynamics (mainly the environments and conditions of the flow and the interaction between the fluids and between the fluids and the porous media); and fourth, the influence of the effective stress. It is these four factors that determine the patterns of nonlinear flows in ultralow-permeability reservoirs.

2.1 STRESS SENSITIVITY OF THE ULTRALOW-PERMEABILITY RESERVOIR

With a complex internal structure, reservoir rocks are typical porous media consisting of solid units (i.e., matrix grains) of various shapes and sizes, among which lie pore spaces in extremely complex shapes. These pores, either interconnected or disconnected, are all filled with one or several fluids. The stress in the porous media is complicated by their unique material structures, which usually bear both external and internal stresses. Variations of the stress states are usually the immediate cause of the deformation of the porous media in petroleum reservoirs.

2.1.1 Features of Stress Sensitivity

The permeability of rocks reflects their overall conductivity and determines the fluid flow conditions. In the process of oil and gas field development, the pressure within reservoir rocks keeps changing. That is to say, the reservoir pressure gradually drops with the progress of development, thus increasing the effective pressure from the overlying rocks. This may compress the reservoir rocks and close the minute channels in them, thus decreasing the reservoir permeability; changes in permeability will inevitably affect the capacity of underground seepage, and then the productivity of oil and gas wells.

The compression of pore spaces in reservoir rocks has an effect on petroleum production in two aspects. One is a positive effect in that the elastic energy released from such compression serves as a driving force for the flow of oil and gas within the pores, and the other is a negative effect in that the lowered permeability caused by the narrowing of pore channels will increase the flow resistance and thus lead to output decline.

The permeability of low- and ultralow-permeability rocks, especially of those under abnormally high pressure, is more sensitive to pressure variations than that of medium- and high-permeability rocks. This is closely related to the structure and matrix features of the rocks and the microscopic features of seepage flows in the low-permeability reservoir.

2.1.1.1 Features of Pore Structures

The pore structures of high- and low-permeability rocks are quite different from each other. High-permeability reservoirs usually contain large pores with thick throats, whose volume accounts for a high proportion of the total pore volume. On the other hand, low- and ultralow-permeability reservoirs often have small pores with thin throats, and the volume of large pores is a low proportion of the total pore volume. Figures. 2-1 and 2-2 present curves of the capillary forces and distribution of pore sizes for

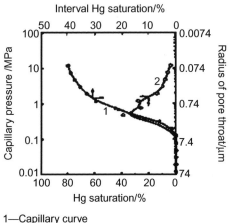

1—Capillary curve
2—Curve of pore size distribution

FIGURE 2-1　The capillary curve of the ultralow-permeability rock ($K = 1.94 \times 10^{-3} \, \mu m^2$).

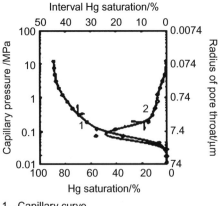

1—Capillary curve
2—Curve of pore size distribution

FIGURE 2-2　The capillary curve of high-permeability rock ($K = 127.24 \times 10^{-3} \, \mu m^2$).

reservoirs of ultralow and high permeability, respectively, from which it can be seen that the effective pores in ultralow-permeability cores are composed of those with small geometric absolute values. Large pore channels serve as the major contributors to the core permeability. When the effective pressure increases, the core permeability tends to decline significantly once the pore channel is compressed and undergoes small changes. On the contrary, in the high-permeability core where the effective porosity is one order of magnitude higher than that in the low-permeability core, small changes in pores have little effect on rock permeability. According to the capillary flow model established by Kozeny (1927), the reservoir permeability is proportional to the fourth power of the average radius of the capillary. Small changes (e.g., decrease) in the average radius of the capillary will cause significant changes (e.g., decrease) in the permeability. Therefore, changes in the pore structure in the ultralow-permeability reservoir are a cause of permeability decline as the confining pressure increases.

2.1.1.2 Features of Matrix Structures

The matrix structure in ultralow-permeability reservoirs is characterized by small matrix grains with very large specific surfaces. Statistical analysis of reservoir features shows that these reservoir rocks are dominated by fine sandstone-siltstones, containing a high content of cement and clay, and a large amount of associated authigenic clay. The cementation types are dominated by basement-pore cementation and pore cementation. There are two main features of the matrix structure in the ultralow-permeability reservoir: one is the very fine matrix grains with very large specific surfaces and the other is the special mechanical properties of the reservoir matrix.

The Relationship between Permeability and Matrix Grains

A specific surface is the sum of the surface areas of the matrix grains per unit volume of rock, indicating the dispersion of the rock and related to the distribution and sizes of the radius of matrix grains. The smaller the matrix grains are, the larger the specific surface will be. W. C. Krumbein and G. D. Monk (1942) established an empirical formula about this:

$$K = Cd^2 e^{-1.35a} \tag{2.1}$$

In the formula, C is the constant coefficient, related to the maturity of the rock, d is average diameter of matrix grains, μm is the standard deviation of matrix grains, and K is the permeability, $10^{-3} \mu m^2$.

Equation (2.1) shows that the rock permeability is proportional to the square of the average diameter of matrix grains and inversely proportional to grain sorting. Under the confining pressure, the matrix grains tend to become smaller, which means the rock permeability tends to decline.

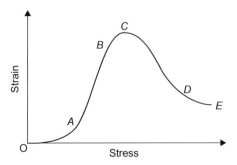

FIGURE 2-3   A typical rock strain-stress curve.

Mechanical Properties of the Reservoir Matrix Grains

A force acting upon the rock may easily transform the matrix structure, thus changing the pore structure. The changes in the matrix structure are usually expressed through the stress-strain curve (Fig. 2-3), which shows that rock containing a certain amount of cement, when stressed, will change first, because the cement is weaker than the matrix grains. The slightly upward OA segment in the curve shows that under confining pressure, the cement structure experiences a strain and is deformed, with the spaces among the grains shrinking, thus narrowing the pore throats. The deformation of this segment has the feature of soft plastic deformation, which means the compacted and deformed cement may not be recovered to its original state even when the confining pressure is restored. The gradient that is reflected in the curves of confining pressure and permeability and the curves of confining pressure and circulation permeability indicate high rates of permeability decline and low permeability recovery rates even after the confining pressure is released.

As the confining pressure increases, the cement connecting the grains is further compacted until its intensity and the space and extent of its deformation approach those of the matrix grains. In Fig. 2-3, the slope of segment AB in the curve is a constant or near a constant, indicating elastic deformation, which can be recovered after the confining pressure is released.

2.1.1.3 Features of Microfractures

In general, a tight sandstone reservoir and its low permeability may lead to a high probability and intensity of fractures and microfractures. The OA and AB segments in Fig. 2-3 can be described this way: the OA segment is at the "work hardening" stage when the curve goes upward, indicating that with the increase of the stress, the speed of the strain growth decreases as if the rock becomes harder with the increase of stress (work). From a microscopic viewpoint the bending of the OA segment is caused by the closure of micropores and microfractures in the rocks under stress. The AB segment is at the linear elastic stage, with its slope (i.e., the effective Young's modulus of rock) determined by the elastic constants of solid material and pores contained in the rock. Assume the pores can be classified into two types: soft pores, which are easily deformed, and hard pores, which have a certain elasticity. When the confining pressure is low, the soft pores begin to deform, shrink, or even close. So during this period the rock permeability decreases significantly and is diffi-cult to recover. As the confining pressure further increases, the deformation of soft pores almost finishes, and the remaining elastic pores may undergo slight elastic deformation only under the confining pressure. Therefore, at this stage the rock permeability decreases linearly as the confining pressure increases and is recovered as it is released.

2.1.1.4 Microscopic Features of Seepage Flows

The permeabilities in low- and ultralow-permeability cores are not only more stress sensitive than those in medium- and high-permeability cores, but are also related to the microscopic features of seepage flows in the low-permeability rock. As a functional parameter, permeability involves both the pore structure and size in the rock on the one hand and the fluids on the other. The fluids within the pore channels will interact with the rock, and have an impact on permeability. Especially for ultralow-permeability cores, the pore system is basically composed of small channels with a very large specific sur-face, which increases the adsorption capacity of the pore surface to fluids and the impact of the fluid boundary layer in the pore. As a threshold pressure gra-dient (TPG) exists when fluids flow in rocks, once the driving pressure is lower than the TPG of some pore channels, these channels will fail to let fluids flow through. When the rock bears greater forces, the channels will shrink, requiring a higher TPG. This causes more small channels to lose their flow capacity, and hence greater reductions in permeability of tight rocks. This makes the permeability of ultralow-permeability cores more sensitive to stress than that of medium- and high-permeability cores. The strength of this effect is related to the composition (e.g., resin, asphaltene content, salinity, etc.) and the PVT(Pressure-Volume-Temperature) state of the fluids. Tests using nitro-gen and standard saline separately to measure the stress sensitivity of the fluids as they flow in the rock show that the value of the stress sensitivity tested with standard saline is higher than that with nitrogen (Fig. 2-4).

2.1.2 Methods and Results of Stress Sensitivity Experiments with Ultralow-Permeability Rocks

2.1.2.1 Measurement of Variations of Porosity and Permeability along with Changes in the Effective Overburden Pressure

Experiment plans and procedures are designed in line with Chinese Petroleum Industry Standard SY5336-88, using the high-temperature and

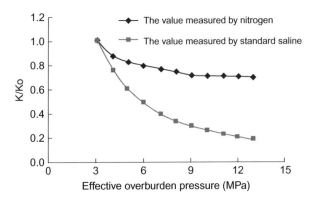

FIGURE 2-4   The stress sensitivity of the rock measured with different fluids.

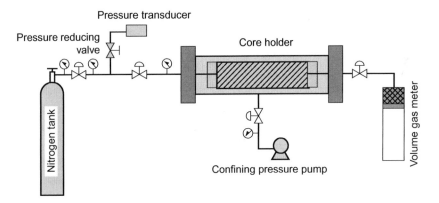

FIGURE 2-5   Flowchart of stress sensitivity experiment with high-pressure porosity and permeability meter.

high-pressure porosity and permeability meter CMS-300 as the experiment instrument, with the experiment flow shown in Fig. 2-5. The selected core sample is extracted with 1:3 alcohol and benzene in a Soxhlet extractor for seven days. When the solution from the siphon becomes colorless and transparent, the sample core is taken out and baked for 48 hours at a temperature of 105°C in the incubator after the organic solvents in the core volatilize completely.

Measurement of variations of porosity and pore compressibility along with changes in the effective overburden pressure

The device that measures compressibility is composed of a core holder, a confining pressure pump, a pore pressure pump, and a thermostat. The core holder is of the pseudoternary-axis pressure type. A simulation of the production process by keeping the overburden (confining) pressure constant while gradually reducing the pore pressure may obtain a compressibility closer to the actual situation.

In the experiment, the confining pressure pump and the pore pressure pump both work at a maximum working pressure of 70 MPa, but with a volume measuring precision of 0.01 ml and 0.001 ml, respectively. The entire experiment device consists of the confining pressure and pore pressure systems, and is set up using the following procedures:

- Calibration of the compressibility of the pore pressure system, i.e., measurement of the $p \sim V$ relationship in the system without loading the sample core.
- Installation of the sample holder without loading the sample but putting confining pressure on the holder.
- Vacuumization of the sample holder so that the pore pressure system is full of water with a certain volume V_w.
- Calibration of the pore pressure system: a. fluid (water) compression; b. pipeline expansion under pressure; and c. deformation of the inner part, e.g., the seal ring of the pump.

To measure the pore volume in the sample:

1. Load the rock sample into the holder and vacuumize it.
2. Inject water of the same volume V_w into the pore pressure system.
3. Establish formation pressure conditions. Maintain a certain pressure difference and simultaneously increase the confining pressure and pore pressure until it reaches a confining pressure of 70 MPa;
4. Keep the pressure stable for a certain period of time.
5. Reduce the pore pressure gradually and measure the $p \sim V$ total relationship.

Calculating Variations of the Pore Volume

Calculating the total volume variations under different pressures, according to the $p \sim V_t$ relationship, i.e., under pressure p the variation of the sample pore volume is $\Delta V_p = \Delta V_t - \Delta V_{pump}$.

The pore compressibility calculation formula is

$$C_p = \frac{1}{V_p} \cdot \frac{\Delta V_p}{\Delta p} \tag{2.2}$$

where
C_p—rock pore compressibility, MPa^{-1}
V_p—rock pore volume, m^3
ΔV_p—rock pore volume changes in value, m^3
Δp—pore fluid pressure changes in value, MPa

Measurement of the Relationship between Permeability Variations and Effective Overburden Pressure Variations

Different fluids can be used to measure the relationship between permeability variations and effective overburden pressure variations. At present, experimenters often use nitrogen as the displacing fluid. In Fig. 2-5, the experiment fluid is high-purity nitrogen. By regressing the apparent permeabilities measured under $4 \sim 10$ different displacing pressures, a Klinkenberg permeability can be obtained, thus eliminating the influence of the slippage effect. The displacing pressure from the nitrogen bottle is controlled through the pressure reducing valve, which keeps the pressure between $0.1 \sim 0.4$ MPa at the core inlet (atmospheric pressure at the exit), thus avoiding an effective stress imbalance between the two ends of the core caused by excessive displacing pressure difference. Through changing the effective stress on the core by changing the confining pressure, the effective stress changes from low to high. For each core, the Klinkenberg permeabilities corresponding to $8 \sim 10$ different effective stresses are measured. Then the effective stress is reduced to measure $5 \sim 8$ Klinkenberg permeabilities corresponding to different effective stresses from high to low.

Experiment method and principle:

$$\frac{dp}{dx} = \frac{\mu_g v}{1 + \dfrac{b}{p}} \times \frac{1}{K_\infty}$$ (2.3)

where

dp/dx—the pressure gradient in the flow direction, MPa/cm

μ_g—gas viscosity, mPa s

v—flow rate, cm^3/s

b—gas slippage factor, MPa

$\bar{p}$—average pore pressure, MPa

K_∞—Klinkenberg permeability, μm^2

2.1.2.2 Experiment Methods for Core Elastic-Plastic Deformation Studies

The triaxial test system for rock mechanics is used as the experiment apparatus; this can be used to study the mechanical behavior of rocks under simulated conditions of reservoir temperature, stress, and pressure.

The system includes the following sections:

- Loading frame and axial pressure control system
- Confining pressure control system

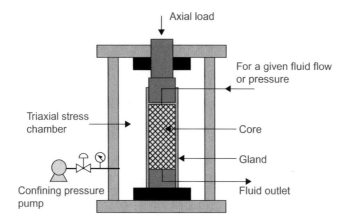

FIGURE 2-6   Diagram of the triaxial stress testing system.

- Pore pressure control system
- Computer data acquisition and control system

  As shown in Fig. 2-6, the main test indicators of this system include:

- Maximum axial load: 1400 KN
- Maximum confining pressure: 140 MPa
- Maximum pore pressure: 100 MPa
- Size of the test rock sample: Ø25 × 50 mm, Ø50 × 100 mm
- Maximum temperature: 200
- System precision: with an error of 0.0005

  While the confining pressure in Fig. 2-6 is used to simulate the horizontal stress in the formation, the axial stress and the pore pressure are used to simulate the overburden pressure and the reservoir pressure, respectively.

  The experiment uses the following procedure:

1. Test the Klinkenberg permeability of the core with nitrogen gas (with a confining pressure of 2.0 MPa).
2. Establish the irreducible water saturation of the core.
3. Put the heat-shrinkable tube over the core to heat the rock sample with a hair dryer until the tube shrinks.
4. Fix axial and radial displacement sensors onto the core and place it into the triaxial stress chamber.
5. Fill the triaxial chamber with oil and increase the confining pressure to the target value at a fixed speed.
6. Add axial stress at a fixed speed until the core is damaged, in the process of which the computer will automatically record the stress and strain data of the core.

2.1.2.3 Results from Stress Sensitivity Test with Ultralow-Permeability Rocks

Influence of Effective Overburden-Pressure Variations on Porosity

Figures 2-7(a) and (b) are diagrams to present how the porosity ($Ø_i/Ø_o$) changes with the effective stress variations in the two ultralow-permeability cores, in which $Ø_o$ is the initial porosity, i.e., effective porosity values measured under an effective overburden pressure of 2.0 MPa, and $Ø_i$ is the value of porosity measured under any effective overburden pressure P. It can be seen from the figures that the porosity curves almost overlap in the processes of pressure boosting and reduction. This means that the final irreversible porosity reduces at a very low rate, i.e., below 2%, and such losses mainly occur when the effective overburden pressure changes between 0.0 MPa and 20 MPa. It can therefore be concluded that under an effective overburden pressure in this range the core undergoes microplastic deformation and when the effective overburden pressure is higher than 20 MPa the core undergoes elastic deformation.

Figures. 2-8 and 2-9 are drawn with the data thus obtained. Note that the core porosity changes only slightly with varying stress, and the loss rate of

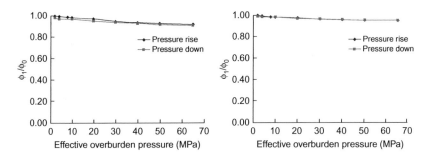

FIGURE 2-7 Diagrams of porosity changes with variations of effective overburden pressure.

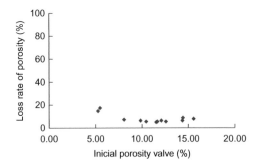

FIGURE 2-8 The relationship between initial core porosity.

porosity is related to the effective stress only and not to the initial porosity or permeability.

Influence of Effective Overburden Pressure Variations on Pore Compressibility

Figures. 2-10(a) and (b) are diagrams that show how the pore compressibility (C_{pi}/C_{p0}) changes with the effective overburden pressure in two ultralow-permeability cores; observe that the pore compressibility changes significantly along with variations of effective overburden pressure. For example, when the effective overburden pressure reaches 65.0 MPa, the pore compressibility loss rate exceeds 80% (Figs. 2-11 and 2-12). It can also be seen that at the initial stage when the effective overburden pressure increases, the pore compressibility reduces rapidly, but after the effective overburden pressure exceeds 20.0 MPa, the reduction of pore compressibility slows down, presenting a linear relationship with the pressure. A qualitative explanation for this phenomenon is that, at the initial stage, the pore

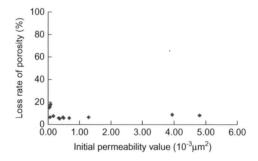

FIGURE 2-9   The relationship between initial core and the loss rate of porosity under pressure permeability and the loss rate of porosity under pressure.

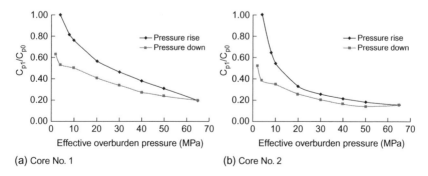

FIGURE 2-10   Graphs that demonstrate how changes in effective overburden pressure affect pore compressibility.

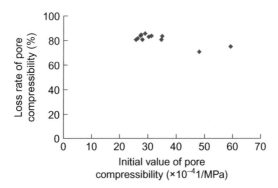

FIGURE 2-11 The relationship between initial pore compressibility and the loss rate of compression factor after pressed.

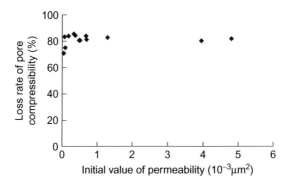

FIGURE 2-12 The relationship between initial factor after pressed permeability and the loss rate of compression factor after pressed.

compressibility changes significantly due to the soft plastic deformation of mud and cement and the closing of microcracks within the rock. As the effective overburden pressure further increases, the rock becomes tighter, which increases the contact area between grains, so that the rock becomes increasingly tight and more difficult to compress. Therefore, when analyzing wide-ranging variations of formation pressure, the rock compressibility cannot be regarded as a constant. In other words, a nonlinear relationship should be used to replace the linear relationship in Hooke's Law.

Figures 2-10(a) and (b) also show that in the pressure release process, the pore compressibility undergoes poor recovery, which suggests that in the development of ultralow-permeability reservoirs, if formation pressure is restored through shut-in, the lost reservoir compressibility cannot be restored to its initial state even after the formation pressure is recovered.

Pore compressibility is an important parameter in the production of reservoirs because the deformation of rocks provides the energy for oil and gas production. Studies have shown that changes in pore compressibility have an effect on features of reservoir development and the ultimate recovery. Most researchers focus only on the sensitivity of permeability when studying stress sensitivity and fluid-rock interaction in ultralow-permeability reservoirs, but it is worth looking at the stress sensitivity of pore compressibility as well.

Influence of Effective Overburden Pressure Variations on Permeability

Figures 2-13(a) through (h) present how the permeabilities at different levels in ultralow-permeability cores are affected by variations of the effective overburden pressure. The Klinkenberg permeability K_∞ is encoded in the graph as K. K_0 refers to the initial permeability, i.e., the Klinkenberg permeability obtained under a pressure of 2.0 MPa. From the experiment results presented in Fig. 2-13 we can draw the following conclusions:

- First, the lower the initial permeability is, the greater the reductions in the permeability as the effective overburden pressure increases, much greater than porosity reductions; when the effective overburden pressure decreases, the permeability cannot be fully restored, and the lower the initial permeability, the greater the irreversible losses of permeability will be.
- Second, the greatest reductions and irreversible losses of permeability occur when the effective overburden pressure ranges between 0.0 Mpa and 20 Mpa; when the effective overburden pressure is greater than 20 MPa, the permeability recovery is better.
- Third, in different cores, the routes of permeability changes with variations of effective overburden pressure are different from each other. No matter how the relationship between permeability and effective overburden pressure is established (e.g., through exponential, power-type, or binomial formulas), the coefficients of different cores may vary from each other. Since each formula involves at least two coefficients, it is difficult to obtain a uniform formula for expressing the coefficients.

Influence of Multirounded Pressure Boosting and Releasing on Permeability

Figure 2-14 shows the test results of pressure ups and downs from the two sample cores. It can be seen that, in each round of pressure ups and downs, the cores suffer different rates of irreversible losses in permeability, the highest in the first round. Similarly to a single round, the lower the initial permeability is, the greater the core's losses are in the pressure rising process and the greater the irreversible losses are in the process of pressure release.

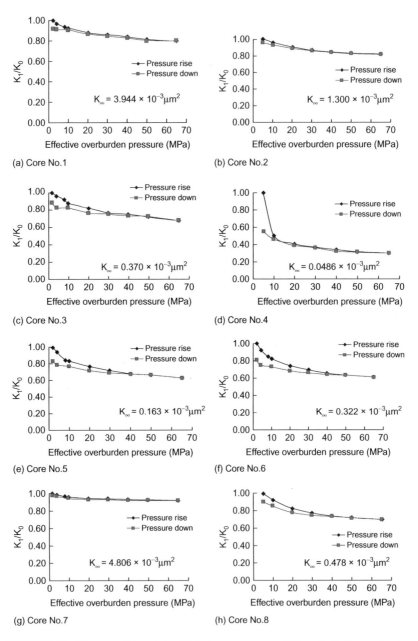

FIGURE 2-13 Diagrams of the relationship between core permeability and effective overburden pressure.

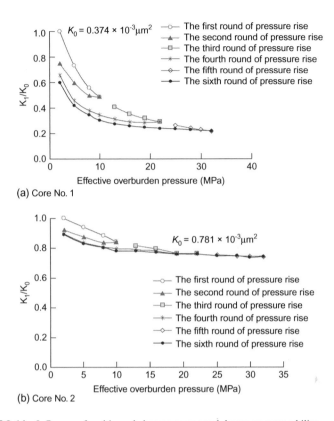

(a) Core No. 1

(b) Core No. 2

FIGURE 2-14 Influence of multirounded pressure ups and downs on permeability.

At the same time, the permeability reduces greatly as the pressure rises when the effective overburden pressure is low. For a relatively high-permeability core under a higher effective overburden pressure, the permeability does not change significantly, with its routes almost the same.

Test Results of Elastic-Plastic Deformation of Ultralow-Permeability Cores

Figure 2-15 presents the relationships between axial stress and strain and between confining pressure and volumetric strain in the ultralow-permeability core. It can be seen that at the initial stage of imposing confining pressure, the core undergoes significant deformation, which is mainly plastic. As the pressure increases, the deformation gradually shifts to a stable elastic deformation stage. The small circle in each part of the figure marks the turning point from plastic to elastic deformation, corresponding to a pressure of about 20 Mpa, which is consistent with the pattern

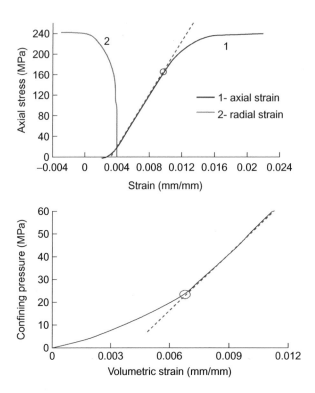

FIGURE 2-15   Diagrams of stress-strain relationship in core D-3.

of porosity variations described earlier. The elastic-plastic deformation occurring at the initial stage of pressure rise (0.0 ~ 20 MPa) is partly caused by the plastic deformation of the mud and cement in the cores, and partly by the closeup of the fine throats and microfine cracks in the rocks.

(a) Diagram of axial stress-strain relationship
(b) Diagram of confining pressure-volumetric strain relationship

The following conclusions can be drawn from a comprehensive analysis of the results from experiments for stress sensitivity and elastic-plastic deformation

- The deformation in the whole process of increasing effective overburden pressure occurs in the order of plastic → elastic → plastic deformation.
- Under a pressure of 0.0−20.0 Mpa, the test cores undergo elastic-plastic deformation, causing unrecoverable losses of porosity and permeability, with those of the latter especially serious. Under a pressure of over 20.0 Mpa, the test cores undergo elastic deformation. Further increasing the confining pressure to the elastic limit σ_e of the core may cause it to enter into the stage of elastic-plastic deformation, and finally, the stage

of plastic deformation. In other words, the elastic deformation stage corresponds to a pressure of $20.0\,\text{MPa} \to \sigma_e$. Because of the differences in rock composition and structure, the σ_e of the core is also different and its value tends to increase as the confining pressure rises.

2.1.2.4 Defining the Stress-Sensitivity Coefficient

Mathematical regression of the relationship between Klinkenberg permeability and effective overburden pressure of dozens of low-permeability cores reveals that of all formulas, the power-type formula produces the highest correlation coefficients (mostly above 0.99), followed by the quadratic polynomial (with lower correlation), while the exponential formula produces the lowest correlation, which shows that the conventional exponential formula method is not fit for ultralow-permeability cores, or in other words, not the most accurate, because of deformation in them.

The prerequisite for an exponential formula that defines the permeability modulus α_K as a constant is based on the hypothesis that the core undergoes elastic deformation under stress. However, an ultralow-permeability core, which contains a high content of cement and clay and well-developed micro-cracks, will undergo plastic deformation first rather than elastic deformation as the overburden pressure increases.

In search of a more reasonable expression for the relationship between permeability and effective overburden pressure, Cheng LinSong and Luo RuiLan of the China University of Petroleum created a new definition for the stress-sensitivity coefficient.

After dimensionless treatment of permeability and effective overburden pressure of the test core, a power-type formula is obtained between the permeability and the effective overburden pressure:

$$\frac{K}{K_0} = a \left(\frac{\sigma_{eff}}{\sigma_{eff0}} \right)^{-b} \tag{2.4}$$

in which the expression for effective overburden pressure σ_{eff} is

$$\sigma_{eff} = \sigma_v - \alpha p \tag{2.5}$$

where

σ_v—overburden pressure, the confining pressure in the experiment while the overburden pressure under reservoir conditions, MPa

p—pore pressure, MPa

α—coefficient of effective stress, ranging from 0.9 to 0.98 for the ultralow-permeability sandstone cores, and approaching 1.0 for reservoirs with fractures or fracture-like pores

When $\sigma_{eff} = \sigma_{eff0}$, $K = K_0$, from which we get that the value of α is 1 in Eq. (2.4).

Then Eq. (2.4) becomes $K/K_0 = (\sigma_{eff}/\sigma_{eff0})^{-b}$.
Using a common logarithm on both sides, we obtain

$$\lg \frac{K}{K_0} = -b\lg \frac{\sigma_{eff}}{\sigma_{eff0}} \tag{2.6}$$

Eq. (2.6) is a straight line through point (1,1), with a slope of $-b$ in double logarithmic coordinates.

$$\frac{K}{K_0} \sim \frac{\sigma_{eff}}{\sigma_{eff0}}$$

Here, a new stress-sensitivity coefficient is defined as

$$S = \lg \frac{K}{K_0} \bigg/ \lg \frac{\sigma_{eff}}{\sigma_{eff0}} \tag{2.7}$$

Thus it is easy to get the stress-sensitivity coefficient S through the fit-of-power relationship of $(K/K_0) \sim (\sigma_{eff}/\sigma_{eff0})$, which is a negative power exponent. Simple in form, this formula has a high degree of correlation with the experiment data. The advantage of this definition for the stress-sensitivity coefficient lies in its uniqueness, i.e., each sample core corresponds to one coefficient, whose value is independent from experiment data points and from the maximum confining pressure.

Figure 2-16 shows the relation curves of cores at different permeable levels.

Regression analysis of experiment data of dozens of tight cores turns out a curve of the relation between the stress-sensitivity coefficient S and the initial permeability K_0, as shown in Fig. 2-17(a), which in double logarithmic coordinates appear as a linear relationship, as shown in Fig. 2-17(b).

It can be seen from Fig. 2-17 that, when the core permeabilities are less than 1.0×10^{-3} μm^2, the stress-sensitivity coefficient increases dramatically,

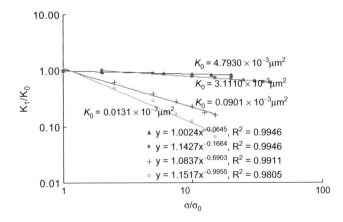

FIGURE 2-16 Diagrams of the relationship between the core permeability and the effective overburden pressure.

which means that stress sensitivity has a significant effect on cores from ultralow-permeability and tight reservoirs. From the regression equation in Fig. 2-17, we can see that the stress-sensitivity coefficient of tight, deformable media has a power relation with the initial permeability, which can be expressed as

$$S = cK_0^{-n} \qquad (2.8)$$

For different reservoirs, the coefficients (c, n) have different values, which should be measured through experiments to obtain more accurate values. In Fig. 2-17, $c = 0.1805$ and $n = 0.432$ are the results for the low-permeability reservoirs in Changqing.

From Eqs. (2.7) and (2.8), we can derive the relationship between the random initial permeability and the effective overburden pressure as follows:

$$\frac{K}{K_0} = \left(\frac{\sigma_{eff}}{\sigma_{eff0}}\right)^{-S} = \left(\frac{\sigma_v - \alpha p}{\sigma_{eff0}}\right)^{-S} = \left(\frac{\sigma_v - \alpha p}{\sigma_{eff0}}\right)^{-(cK_0^{-n})} \qquad (2.9)$$

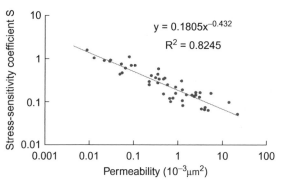

(a) Double-logarithmic coordinates

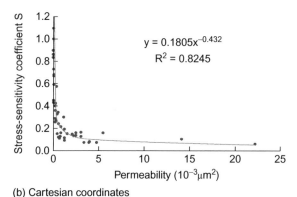

(b) Cartesian coordinates

FIGURE 2-17   Diagrams of the relationship between the initial permeability and the stress-sensitivity coefficient.

Eq. (2.9) shows that, so long as the coefficients c and n are determined (i.e., the formula for the stress sensitivity factor S is established), the relationship between the permeability and effective overburden pressure can be expressed. Then it is easy to calculate the dynamic changes of any permeabilities in heterogeneous reservoirs, thus laying a foundation for flow modeling and numerical simulation of ultralow-permeability reservoirs with deformable media. Eq. (2.9) can also be used to convert the core permeability obtained under low confining pressure on the ground into the permeability under reservoir conditions so that it can be of help for a correct evaluation of reservoir productivity, and thus of practical engineering applicability.

2.1.3 Theoretical Interpretation for the Stress Sensitivity of Reservoir Permeability

Patterns or rules of permeability variations have always been a focus in the study of the percolation mechanism of fluid-solid coupling in reservoirs with deformable media. Many lab studies about how permeability changes with stress variations and established relevant models have been completed. But no consistent conclusions have been arrived at. Some hold that the rates of permeability variations along with effective stress are closely related to the initial permeability: the lower the initial permeability is, the more sensitive it is to the stress. But others argue that the permeability variation isindependent from the initial permeability value.

Suppose the porous media and their flow space are composed of parallel capillary bundles. Then the capillary volume is equivalent to the pore volume in the core while the percolation capacity of capillary is equivalent to the core permeability. Capillary models apply to well-cemented sandstones.

Changes in the effective stress on the core mean that the stress on capillaries has changed, which will inevitably lead to corresponding changes in the size of capillaries and thus their percolation ability. Let's take one capillary from the porous media for analysis, as shown in Fig. 2-18. The capillary has an inner radius of a, and an outer radius of b, under an internal and external pressure of p_a and p_b, respectively. The ratio of the single-capillary volume to the unit volume is called core porosity. Suppose the capillary is a kind of elastic medium. Changes in internal and external stress on the capillary will cause its elastic deformation. According to the thick-wall cylinder theory in elastic mechanics, the relation of capillary strain to stress changes can be expressed as

$$u_r = \frac{1-\nu}{E} \cdot \frac{a^2 p_a - b^2 p_b}{b^2 - a^2} r - \frac{1+\nu}{E} \frac{a^2 b^2 (p_b - p_a)}{(b^2 - a^2)} \cdot \frac{1}{r} \qquad (2.10)$$

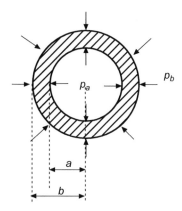

FIGURE 2-18   Capillary diagram.

Suppose there are two capillaries, i.e., 1 and 2, one thicker than the other. Their respective inner and outer radiuses are a_1, b_1, a_2, and b_2. They have the same porosity but different internal diameters, i.e., $a_1 > a_2$, and $(a_1^2/b_1^2) = (a_2^2/b_2^2)$. Both the capillaries will deform under the stress of the same strength p_a and p_b. The inner and outer radiuses will change respectively as in the following expressions:

$$u_{a1} = \frac{a_1}{E} \cdot \frac{1-\nu}{1 - \frac{a_1^2}{b_1^2}} \left[\left(\frac{a_1^2}{b_1^2} - 1 \right) p_b - \left(\frac{a_1^2}{b_1^2} + 1 \right) (p_b - p_a) \right] \qquad (2.11)$$

$$u_{b1} = \frac{b_1}{E} \cdot \frac{1-\nu}{\frac{b_1^2}{a_1^2} - 1} \left[\left(1 - \frac{b_1^2}{a_1^2} \right) p_a - \left(\frac{b_1^2}{a_1^2} + 1 \right) (p_b - p_a) \right] \qquad (2.12)$$

$$u_{a2} = \frac{a_2}{E} \cdot \frac{1-\nu}{1 - \frac{a_2^2}{b_2^2}} \left[\left(\frac{a_2^2}{b_2^2} - 1 \right) p_b - \left(\frac{a_2^2}{b_2^2} + 1 \right) (p_b - p_a) \right] \qquad (2.13)$$

$$u_{b2} = \frac{b_2}{E} \cdot \frac{1-\nu}{\frac{b_2^2}{a_2^2} - 1} \left[\left(1 - \frac{b_2^2}{a_2^2} \right) p_a - \left(\frac{b_2^2}{a_2^2} + 1 \right) (p_b - p_a) \right] \qquad (2.14)$$

With the help of $(a_1^2/b_1^2) = (a_2^2/b_2^2)$ and Eqs. (2.11) through (2.14), the relationship of strains in the two capillaries can be expressed as

$$\frac{u_{a1}}{a_1} = \frac{u_{a2}}{a_2} \qquad (2.15)$$

$$\frac{u_{b1}}{b_1} = \frac{u_{b2}}{b_2} \qquad (2.16)$$

After the strain occurs, the degree of permeability change in the two capillaries (one courser than the other) can be expressed as

$$\frac{K_1}{K_{1i}} = \frac{\phi_1 \cdot (a_1 - u_{a1})^2}{\phi_{1i} \cdot a_1^2} = \left(1 - \frac{u_{a1}}{a_1}\right)^4 \bigg/ \left(1 - \frac{u_{b1}}{b_1}\right)^2 \qquad (2.17)$$

$$\frac{K_2}{K_{2i}} = \frac{\phi_2 \cdot (a_2 - u_{a2})^2}{\phi_{2i} \cdot a_2^2} = \left(1 - \frac{u_{a2}}{a_2}\right)^4 \bigg/ \left(1 - \frac{u_{b2}}{b_2}\right)^2 \qquad (2.18)$$

From Eqs. (2.15) to (2.18), we can infer that $(K_1/K_{1i}) = (K_2/K_{2i})$, i.e., the two different capillaries with different initial permeabilities undergo the same degree of change under stresses of the same strength.

Therefore, it can be concluded that, in theory, the permeability of different cores will have the same rates of reduction under the same effective stress, which will also lead to the same rates of output decline. There is no proof that the stress sensitivity of ultralow-permeability reservoirs is stronger than that of the medium- or high-permeability ones.

The reason why ultralow-permeability cores are more stress sensitive than medium- and high-permeability ones lies in the TPG for seepage flows. Suppose the rock flow channels consist of some microcapillaries with different radiuses. Rheological studies show that the TPG is inversely proportional to the capillary radius. In other words, a shorter capillary radius involves a higher TPG. Theoretically speaking, fluids flowing in porous media require a TPG at different levels. Different sizes of the pore channels lead to different effects of the solid-liquid interface and of the oil boundary layer. Therefore, different pores have different TPGs.

The pore system in ultralow-permeability reservoirs is mainly composed of small channels that have very large specific surfaces and whose fluid boundary layers have a great influence. As the stress increases, the flow channels narrow down while the TPG increases. Then the smallest channels begin to lose seepage capability first. With the stress further increasing, more and more small channels lose their seepage capability, thus causing the stress sensitivity of permeability. However, for medium- or high-permeability reservoirs, whose pore system is dominated by large channels, the increase of stress can only cause their flow capacity to reduce to a certain degree instead of complete loss, indicating that the degree of dimensionless flow decline is related to the effective stress only.

The existence of TPG makes ultralow-permeability reservoirs more stress sensitive than medium- and high-permeability reservoirs. When the initial permeability is below a certain critical value, a lower permeability is more stress sensitive; when the initial permeability exceeds this value, the degrees of permeability variations corresponding to the stress changes are independent of the initial value of permeability. During the production of ultralow-permeability reservoirs, the formation pressure tends to drop as the oil is taken out, which will cause dramatic reductions of the permeability near the wellbore and thus a substantial decline in per-well output. As a consequence, the cycle of well stimulation (such as fracturing and acidizing) is shortened, which makes the development of these reservoirs more costly and difficult.

2.1.4 Factors That Influence the Stress Sensitivity of Ultralow-Permeability Reservoirs

Deformation of porous media is mainly caused by the changes of relative positions of the point mass within them. Next are the internal factors that influence the stress sensitivity of ultralow-permeability reservoirs.

2.1.4.1 Material Composition

The matrix materials in the deformable porous media are made up of either one single mineral or several minerals. For example, quartz sandstones are mainly composed of quartz ($SiO2$), while greywacke is composed of quartz, feldspar, mica, clay, and other minerals. Due to the different hardness of different mineral constituents, hard constituents are unlikely to deform under external forces. But the softer a constituent is, the more easily it will get deformed. Take greywacke, for example. Quartz has the highest hardness, feldspar is second, and mica and clay are the lowest or the softest. Under external forces, mica and clay may get easily deformed or broken and dislocated, which reduces the pore volume in the porous media and even plugs the pores and throats, thus reducing their porosity and permeability.

2.1.4.2 Grain Types

The units of solid substances exist in three main shapes, namely, linear, planar, and granular. In reality, deformable porous media seldom contain linear and planar unit bodies, but mostly irregular granular ones. Quartz sandstones, for example, are dominated by well-rounded quartz grains. Because of the dominance of granular units, such porous media may be prone to displacement and deformation under external forces. However, this is not absolutely true since medium displacement and deformation also depend on how the unit bodies are contacted and cemented.

2.1.4.3 Grain Contact Modes

The degree of deformation is largely affected by the modes of contact between grains in the deformable reservoir. In general, there are four modes

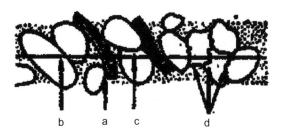

FIGURE 2-19 Modes of contact between grains: a. point contact, b. line contact, c. male and female contact, d. sutural contact.

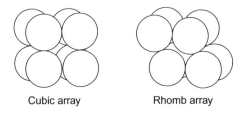

Cubic array   Rhomb array

FIGURE 2-20 Typical array patterns of equivalent-sized spherical particles.

of contact between grains (Fig. 2-19): point contact, line contact, concave-convex (male and female) contact, and sutural contact. Point contact is unstable and easily deformed under external forces, while the other three contacts are more stable and less deformable. In deep reservoirs, rock grains contact each other mainly in the line and concave-convex modes.

2.1.4.4 Patterns of Grain Arrays

The complex shape and size of unit bodies in the deformable medium lead to extremely complex arrays, which for the convenience of research can be illustrated using simplified equivalent-sized spherical particle materials (Fig. 2-20).

Suppose these equivalent-sized spherical particles are arrayed symmetrically. There may be two patterns of array, i.e., cubic array and diamond array. While the cubic array may have a porosity up to 47.6%, it has small coordination numbers, and so is often called a loose array. On the other hand, the diamond array has a porosity of only 26%, but since it has larger coordination numbers, it is often called a compact or tight array. The media in a cubic array are prone to deformation under external forces. The tight, low-permeability rocks, whose porosities are often far below 26%, are dominated by line or interfacial contact between grains, with point contact accounting for only a low percentage. However, such a simplified ideal model may be greatly different from the real situation.

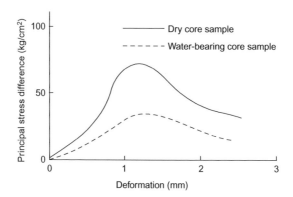

FIGURE 2-21   Humidity's influence on deformation.

2.1.4.5 Modes of Cementation

Not all the spaces between rock grains are porous because various types of cementation occur in the vast processes of deposition and diagenesis. According to sedimentary petrology, the relationship between the grains and the cements fall into four types: base cementation, pore cementation, contact cementation, and poikilitic cementation. From the viewpoint of cement composition, cementation can also be divided into argillaceous cementation, calcareous cementation, siliceous cementation, and so on. Cementation enhances the stability within the rocks, which makes it difficult for their grains to deform and move under external forces. However, different types of cementation have great influence on the physical properties of the rocks. For instance, sandstones with calcareous and siliceous cementation may undergo only elastic deformation when the formation pressure changes significantly, while argillaceous sandstones may suffer elastic-plastic deformation under the same conditions.

2.1.4.6 Fluid Types and Features in Pores

The reservoir may contain oil, gas, and/or water. Different fluid types and features, which entail differences in both the volume elastic modulus and the change patterns of fluid pressure in the pores, may have great influence on the deformation of porous media. For example, a higher water saturation (i.e., humidity as used in rock mechanics), may lead to a smaller elastic modulus, and thus require a lower effective stress to produce the same deformation in the rock (Fig. 2-21).

2.2 TPGs IN ULTRALOW-PERMEABILITY RESERVOIRS

Since the small channels for oil, gas, and water cause high flow resistance in ultralow-permeability reservoirs, the seepage flows, no longer following the classic Darcy's Law, become slow nonlinear flows, thus acquiring an important feature that the flowing process involves a TPG.

2.2.1 Features of TPGs in Ultralow-Permeability Reservoirs

2.2.1.1 The Boundary Layer Theory

Crude oil is a multicomponent hydrocarbon compound, also containing some oxygen, sulphur, nitrogen, and metallic compounds. The main components of crude oil include hydrocarbon and nonhydrocarbon compounds and large amounts of colloid asphaltene, all of which, except for the paraffin hydrocarbons, contain different contents of polar materials. Colloid asphaltenes in the crude are rich in polar components. A higher content of polar components may result in a stronger polarity and surface activity of colloid asphaltenes. Therefore, obvious interaction may occur between colloid asphaltenes and rock grains when they meet each other, forming a special liquid layer rich in polar materials on the surface of rock grains. This layer, attracting more heavy oil components and colloid asphaltenes with viscosities and densities much higher than the corresponding values of bulk oil, is then called the boundary oil layer.

2.2.1.2 The Concept of Seepage Fluids

Mechanical properties of seepage fluids are related to the distribution of porous media, the flow properties and movements, and the interaction between them. For a long time, people have considered the fluids existing in the central section of the porous channel the same as those directly contacting the inner surface of the porous media, which is incorrect. Huang Yanzhang *et al.* defined seepage fluids as follows:

- Seepage fluids refer to those existing in vadose zones, including the bulk fluids and the boundary fluids.
- The properties of bulk fluids refer to those which are distributed around the axis of the channels in porous media and whose properties are affected by the boundary conditions (Fig. 2-22).
- Boundary fluids refer to those attached to the inner surface of the pore channels to form a boundary layer, their properties easily affected by the boundary conditions. Properties of boundary fluids follow special change rules. The porous media are filled with fluids, whose components may interact with the molecules on the pore surface. For this reason, the components near the pore surface tend to have a higher density than those far away from the pore surface. The density variations of these components corresponding to the changes in their distance from the pore surface will cause changes in their physicochemical properties. Therefore, properties of seepage fluids follow special change rules on account of the existence and influence of such boundary fluids in vadose environments.

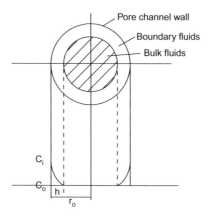

FIGURE 2-22 Distribution of bulk fluids in the channel.

- Fluid properties are determined by the combined properties of bulk fluids, boundary fluids, and porous media and the flow conditions.

It can therefore be concluded that the major cause of TPG in ultralow-permeability reservoirs is the boundary layer in porous media, which creates the problem of non-Darcy flows in such zones because their patterns are rather complicated and thus difficult to deal with.

The concept of seepage fluids enriches the theoretical basis for the study of percolating rules and the percolation theory itself and provides a practical method to accurately describe the percolating process in ultralow-permeability porous media.

2.2.1.3 Viscosity of Crude Oil in the Pore Channels

Crude oil is a multicomponent mixture of hydrocarbons. Its viscosity, besides being influenced by pressure, temperature, and other environmental conditions, mainly depends on the various compositions and the content of components, such as high-molecular hydrocarbons, glial, asphaltene, and metal complexes, which affect not only the average viscosity of oil, but also the distribution of its viscosity in the pore channels. As these substances have a strong interaction with the pore surface, they can be easily adsorbed into the pore wall. Thus, the various components of crude oil in pore channels are distributed in a good order, i.e., the polar materials and heavy components tend to concentrate near the channel walls and decrease gradually in the direction of the pore axis. Such distribution of oil components directly affects the distribution of its viscosity, which is also put into a good order, i.e., the oil viscosity decreases from near the channel walls in the direction of the pore axis. In other words, the pattern of viscosity distribution in pore channels is that the high viscosity of the boundary oil may gradually decrease in the direction from the pore surface to the channel axis, at last replaced by the low viscosity of the bulk oil.

Expression of Oil Viscosity in Pore Channels

As has been analyzed above, the distribution patterns of oil components may cause corresponding changes in oil viscosity distribution. The curve in Fig. 2-23 shows the changes of oil viscosity from the pore surface to pore center r_0. Accordingly, the viscosity of crude oil in the pore can be expressed as

$$\mu = \frac{\int_0^{r_0} f(\mu)2\pi(r_0 - r)dr}{\pi r_0^2} \tag{2.19}$$

where the constitutive equation about viscosity $f(\mu)$ is not easy to resolve and interpret. μ_1 in Fig. 2-23 is the average viscosity of the oil in the boundary layer, μ_2 is the viscosity of bulk crude oil, while h is the thickness of the boundary layer. Then the oil viscosity can be expressed as

$$\mu = A\mu_1 + (1 - A)\mu_2 \tag{2.20}$$

where A is the ratio of boundary oil volume to total oil volume, which can be approximated through the following formula:

$$A = 2\frac{h}{r_0} - \left(\frac{h}{r_0}\right)^2 \tag{2.21}$$

Relationship between A and the Formation Permeability K

From the relationship between the capillary theory model and Darcy's Law, we can derive the relationship between the pore radius of porous media and the formation permeability:

$$r_0 = 0.35\sqrt{K} \tag{2.22}$$

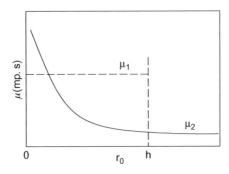

FIGURE 2-23   Changes of oil viscosity in pore channels.

where

r_0—pore radius, μm

K—formation permeability, $10^{-3}\ \mu m^2$

Putting Eq. (2.22) into Eq. (2.21):

$$A = \frac{2h}{0.35\sqrt{K}} - \left(\frac{h}{0.35\sqrt{K}}\right)^2 \tag{2.23}$$

As Eq. (2.23) shows, A is a quadratic function about the relationship between the thickness of the boundary layer and the formation permeability, whose value is closely related to formation permeability and oil properties. In the case of fixed oil properties, A will increase as the permeability decreases. When reaching a certain point, the A value will have a significant influence on the percolating process. This is the basic reason for non-Darcy flow patterns and the existence of TPG in the low-permeability reservoir.

Relationship between Boundary Layer Thickness h and Oil Permeability K

It is widely accepted that the pore radiuses in a reservoir pore system vary in a wide range. But those in low-permeability reservoirs are very short. Experiments show that shorter capillary radiuses may increase the oil boundary layer thickness, while lower permeabilities and shorter pore radiuses may affect heavy oil components more significantly. Therefore, the boundary layer thickness is closely related to oil properties, capillary radiuses (i.e., formation permeability), and the displacing pressure gradient.

The relationship between the boundary layer thickness and the capillary radius can be expressed as follows:

$$h = a - b \ln r \tag{2.24}$$

where

a—coefficient related to oil properties and displacing pressure gradient

b— slope of a curve, approximately 0.0471 for most oil samples

Putting Eq. (2.22) into Eq. (2.24):

$$h = a_0 - 0.0236 \ln K \tag{2.25}$$

where

a_0—coefficient related to oil properties and displacing pressure gradient

As Eq. (2.25) shows, the lower the formation permeability is, the thicker the oil boundary layer will be. Because of the complexity in measuring the boundary layer under experimental conditions, given a permeability and its corresponding boundary layer thickness, Eq. (2.26) can be used to obtain the thickness of the boundary layer for any corresponding permeability under the same experimental conditions within the range of permissible engineering errors:

$$h_2 = h_1 + 0.0236 \ln \frac{K_1}{K_2} \tag{2.26}$$

Relationship between Oil Viscosity μ and Formation Permeability K in Porous Media

Since a_0 in Eq. (2.25) is a coefficient related to oil properties and displacing pressure gradient, we can use the following equation to replace a_0:

$$a_0 = \frac{\beta}{\left(\dfrac{\Delta p}{L}\right)^2} \tag{2.27}$$

where β is a coefficient only associated with the oil properties and changes as oil components vary. But for a certain field, it can be regarded as a constant. Then the thickness of the oil boundary layer in ultralow-permeability porous media can be expressed as

$$h = \frac{\beta}{\left(\dfrac{\Delta p}{L}\right)^2} - 0.0236 \ln K \tag{2.28}$$

Hence, for a formation with a fixed permeability, the pore throat radius and the thickness of the oil boundary layer can be expressed through Eq. (2.22) and Eq. (2.28), respectively. The ratio, namely A, of boundary oil volume to total oil volume can be obtained by combining these two equations with Eq. (2.21). The viscosities of the boundary oil and of the bulk oil can be obtained easily, so we can use Eq. (2.20) to obtain the viscosity of the oil in porous channels.

2.2.1.4 TPG Features in Ultralow-Permeability Reservoirs

Reservoir permeability can reflect the conditions of pore structure to some extent. The pores in ultralow-permeability porous media are characterized by small sizes, fine throats, developed micropores, large specific surfaces, high pore-to-throat ratios and frequent changes of throats. Lower permeabilities may mean shorter average pore radiuses, stronger heterogeneity, higher ratio of micropore volume to total pore volume and higher proportions of boundary fluids in the pore. All this will significantly affect the interaction between liquids and solid interfaces, which will attach the polar materials in the oil to the surface of rock grains. Lower permeability may generate stronger interaction between the liquid and solid interface.

Мархасин И.Л., a former USSR scholar, once pointed out that the boundary layer is formed through the adsorption of active oil components onto the surface of porous media. The composition and properties of the crude oil in the boundary layer, greatly different from those of the bulk oil, have the features of orderly changes of components, structural viscosities, and yielding values. The thickness of the boundary layer is related not only to the properties of crude oil, but also to the pore sizes and driving pressure gradients. For the liquids to flow in the reservoir, the driving pressure gradient must overcome certain additional resistance.

In general, water is a kind of Newtonian fluid, but in very small pores it also follows a non-Newtonian flow model and requires a TPG. This is also true with crude oil. In medium- and high-permeability reservoirs containing light oil, the flow channels for crude oil are not too small, the boundary oil layers are not too thick, the ratio of boundary oil volume to total oil volume is not too high, and the non-Newtonianism of boundary oil has little effect on linear seepage. Thus, people have successfully solved many problems in engineering design and calculation in such reservoirs by using Darcy's Law. However, for ultralow-permeability reservoirs or the heavy oil reservoirs, the effect of non-Newtonianism of boundary crude oil on linear seepage cannot be ignored, for it will change the seepage rules apparently and generate a starting pressure.

Permeability of porous media, as a statistical parameter for evaluation, is the total of many permeabilities of pore channels of various sizes. The pores in high-permeability reservoirs are mainly composed of large pore channels, so it is not easy to monitor the starting pressure as oil or water flows in these pores. Therefore, in our experiments with high-permeability cores, the flow rate and the TPG appear as a straight line through the origin in a Cartesian coordinate system. On the other hand, the pore system of low- or ultralow-permeability cores basically consists of small channels, each having its own TPG for oil and/or water to flow. When the driving pressure exceeds the TPG of a channel, the oil and water begin to flow, thus raising the permeability. With the driving pressure gradient further increasing, more and more channels are involved in the flow, making the core more permeable, with the general permeability enhanced. Thus, in our experiments with ultralow-permeability cores, the flow rate and the pressure gradient in the Cartesian coordinate system is no longer a single straight line, but composed of two parts: an upturned curve and a straight line, which suggests that the permeability increases with the increase of the pressure gradient and reaches a fixed value.

The cross-section of the porous media has a certain transparency, which, from a statistics point of view, is equivalent to the porosity of the media. Because the rock has a very low compressibility, the porosity can be regarded as a constant. Owing to the existence of the oil boundary layer, the actual area available for flow is smaller than the total cross-sectional area of the porous channels, which is also related to the pressure gradient. If the pressure gradient is low, the fluid can only flow in big pores. Only after the pressure gradient reaches a certain value will the small pores begin to work. Therefore, the ratio of the fluids involved in actual flow to the total fluids is called flow saturation, and the ratio of the volume involved in actual flow to the total core volume is called flow porosity. Both of them are functions of the pressure gradient. For the medium- and high-permeability reservoirs containing light oil, as the pressure gradient increases, the porosity can quickly reach a stable value. When it comes to ultralow-permeability or heavy-oil reservoirs, however, things become much more complicated because the seepage rules are changed.

The pore radiuses of the ultralow-permeability reservoirs are very short, mostly less than 1 μm, so the influence of the oil boundary layer is significant, which requires a TPG in the flow process. Various research results indicate that TPG stands in inverse proportion to permeability, i.e., the lower the permeability, the higher the TPG.

In conclusion, the physical account for the existence of TPG can be summed up as follows:

- The fluid seepage velocity is influenced by the size, shape, and distribution of pores. The tight rocks with narrow pore throats, poor connectivity, and low permeability are significant geologic factors of non-Darcy flows. The filtrational resistance greatly reduces the seepage velocity when fluids go through this kind of rock.
- There is always an interfacial interaction between the solid and liquid (gas) phases when fluids percolate through the porous media. On the one hand, surface-active materials in the fluids form an adsorbing layer on the surface of rock grains and stick to the surface of pore throats, either narrowing the pore throats or plugging part or the whole of pores. As a consequence, the permeability drops down sharply and the seepage velocity decreases. On the other hand, the wafers that make up clay have the capacity of absorbing polar molecules in water, which will form a strong hydrated layer on the surface of pores when fluids percolate through the clay. This will also plug the channels. A significant cause of the non-Darcy flow in ultralow-permeability reservoirs is the molecular force on the solid-liquid interface, especially for the seepage of oil and water in place.
- The physical and chemical balance of clay minerals in the reservoir will be broken when they are exposed to low-salinity external fluids (i.e., injected water), which will result in their swelling, dispersion, and migration. The clay solution caused by clay swelling and dispersion of such hydrate swelling is a factor not to be neglected because it will also affect the flow features of injected water.
- Tight rocks, such as shales and mudstones, can imbibe the salt components in water, which, filtrated and precipitated, will plug the pore throats, thus affecting the percolation of fluids.

- The rock frame will be deformed or even broken with the rise of effective stress, dramatically reducing the permeability and porosity, which cannot go back to their original values even when the rock pressure is recovered because permeability is extremely sensitive to pressure.
- The rheological properties of the fluids themselves are also important influential factors.

Through theoretical studies and lab experiments, the process of fluid percolation in ultralow-permeability reservoirs can be described as follows: The molecular forces on the solid surface capture the bound water, thus forming an immobile layer at a low pressure gradient. The thickness of the immobile layer can be expressed as $hs = h0\exp(-c1\mathrm{grad}p)$, which will decrease exponentially as the pressure gradient increases, indicating that the cross-sectional area for the bulk fluids will increase. In other words, the actual cross-section area for flow is smaller than that of the pores in porous media because of the existence of boundary oil layer. Besides, the cross-sectional area for fluids to pass through porous media is related to the pressure gradient. The percentage of moving fluids changes along with the change in the pressure gradient, so the volume of moving fluids is a function of the pressure gradient.

In short, the existence of the boundary oil layer in ultralow-permeability porous media causes the fluids to break Darcy's Law in the percolating process, not only changing the flowing cross-section area, but also reducing the flowing saturation. More seriously, it makes it very difficult for fluids in some small pores to flow. As the driving pressure gradient increases, the thickness of the boundary layer is gradually reduced and eventually stabilized, with the permeability approaching a constant.

2.2.2 TPG Experiments for the Ultralow-Permeability Reservoirs

Fluids in ultralow-permeability reservoirs are quite different from those in high-permeability ones. The most essential and obvious difference is that the former no longer comply with the typical Darcy's Law because they are greatly affected by the solid-wall effect. The "pressure-difference and flow-rate" curve appears as a combination of a curve segment and a quasi-straight line segment, which does not pass through the origin in the Cartesian coordinate system.

2.2.2.1 Experiment Principles

The typical percolation curve for the ultralow-permeability core is a combination of a curve segment and a quasi-straight line segment. In Fig. 2-24, *a* represents the actual TPG, i.e., the TPG for the fluids to flow in the largest pores in porous media, and *b* represents the proposed TPG, i.e., the TPG for the fluids to flow in the smallest pores. Line segment *de* extends forward

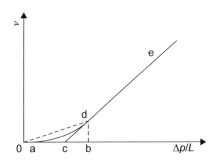

FIGURE 2-24   The typical curve of the flow through an extra-permeability core.

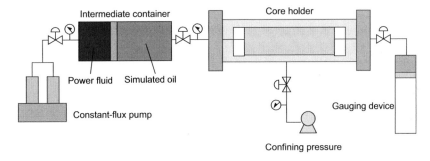

FIGURE 2-25   Experiment procedures.

until it crosses the pressure gradient axis at point *c*, which is the TPG required for the experiment, and stands for the average TPG of the core. Thus the core TPG can be achieved with the method of "pressure difference and flow rate," i.e., by measuring the differential pressure at both ends of the core and the rate of flow that passes through it under this differential pressure.

2.2.2.2 Experiment Procedures

Our experiment follows the basic procedures shown in Fig. 2-25. The kerosene (power fluid) as the displacing medium is pumped by a constant-flux pump into the intermediate container to push the piston toward its other side so that the model oil (or water, i.e., the power fluid) in the container is compressed into the core and flows through it under the differential pressure. After the flow becomes stable, the time taken by a certain volume of liquids to flow through the core is measured.

2.2.2.3 Experiment Methods

The making and washing of the test core and preparations for the experiment are similar to what is done in our stress-sensitivity evaluation experiment. The effective stress on the rock is simulated by the confining pressure to maintain the displacing pressure gradient within the range of real flowing conditions in a reservoir, which is gradually reduced. The experiment procedures can be described as follows:

1. Prepare the core sample and ensure that the equipment is airtight.
2. Vacuumize the sample and fill the intermediate container with the power fluid.
3. Put confining pressure on the core holder to drive the remaining gas out of it.
4. Turn on the constant-flux pump to push the power fluid from the intermediate container into the core under a certain stable pressure.
5. Measure the outflow velocity of the power fluid with a stopwatch and capillary after the flow becomes stable and record it.
6. Adjust the outlet pressure of the pump to allow different outflow velocities.
7. Sort out the data to get some results.
8. Replace the core and power fluid for repeated experiments.

2.2.2.4 Data Processing

The percolating equation taking TPG into account is

$$v = -\frac{K}{\mu}\nabla p\left(1 - \frac{G}{|\nabla p|}\right), |\nabla p| \geq G$$

$$v = 0, |\nabla p| < G$$

(2.29)

i.e., under one-dimensional conditions, the non-Darcy law can be expressed as

$$Q = \frac{K}{\mu}A\left(\frac{\Delta p}{\Delta l} - G\right)$$

(2.30)

The highest displacing pressure gradient $\Delta p/\Delta L$, when $Q = 0$, turns out to be the TPG λ.

Equation (2.30) can be changed into

$$\frac{KA}{\mu}\frac{\Delta p}{\Delta L} - Q = \frac{KA}{\mu}G$$

(2.31)

Obtaining the logarithm for both sides of Eq. (2.31) can produce another equation:

$$\log\left(\frac{KA}{\mu}\frac{\Delta p}{\Delta L} - Q\right) = \log\left(\frac{KA}{\mu}\right) + \log G$$

(2.32)

where $\log G$ and $\log((KA/\mu)(\Delta p/\Delta L) - Q)$ have the relationship of a straight line with a slope of 1 in the Cartesian coordinate system. When $\log[(KA/\mu)(\Delta p/\Delta L) - Q] = 0$, $\log G = -\log(KA/\mu)$, i.e., $G = (\mu/KA)$.

There are various factors affecting the TPG. According to Eq. (2.32), $\log G = -\log(KA/\mu)$, i.e., $G = (\mu/KA)$ when $\log[(KA/\mu)(\Delta p/\Delta L) - Q] = 0$. Therefore, factors affecting the single-phase TPG mainly include properties of the porous media and the fluids.

2.2.2.5 Data Analysis

TPGs of typical low-permeability cores from an oilfield in China are measured respectively by using model oil, formation water, injected water, and distilled water, at a temperature of 21 °C, and with the gas-logging permeabilities ranging between 0.022 and 8.057×10^{-3} μm^2. The fluid parameters are listed in Table 2-1.

The curves in Fig. 2-26 show the relationship between TPG and permeability with different displacing media, suggesting that the TPG measured changes are generally in the same trend in spite of different displacing fluid

TABLE 2-1 Fluid Parameters in the Experiment

Number	Test fluid	Density (g/cm^3)	Viscosity (mPa · s)	Notes
1	Model oil	0.80	1.25	On-the-spot degassed oil + kerosene
2	Formation water	1.03	0.91	Water type $CaCl_2$, salinity 49.35 g/L
3	Injected water	0.98	0.85	Water type Na_2SO_4, salinity 1.709 g/L
4	Distilled water	0.95	0.83	—

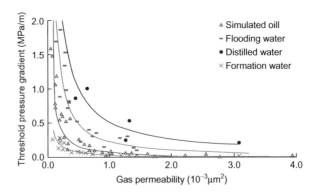

FIGURE 2-26 Relationship between TPG and permeability under different displacing fluids.

properties. The TPG is low and increases slowly when the permeability is higher. With the permeability reduces, the TPG increases. When the permeability is below a certain value, a sharp increase occurs in the TPG. The regression equations of different fluids can be shown through numerical matching, as follows:

For model oil: $G_o = 7.473 \times 10^{-2} \; K^{-1.117} \; R^2 = 0.9118$ (2.33a)

For formation water: $G_{wf} = 3.980 \times 10^{-2} \; K^{-0.935} \; R^2 = 0.9165$ (2.33b)

For injected water: $G_{wi} = 2.306 \times 10^{-1} \; K^{-1.096} \; R^2 = 0.8677$ (2.33c)

For distilled water: $G_{wd} = 5.177 \times 10^{-1} \; K^{-0.952} \; R^2 = 0.8429$ (2.33d)

where

K—permeability, $10^{-3} \; \mu m^2$
G—TPG, MPa/m
Superscripts—o represents the oil phase and w water phase

These four equations show that the TPGs obtained with different test fluids have a relationship of power function with the core permeabilities. While the fluid is percolating in the reservoir, the adsorption between the polar molecules in the model oil and the rock grains may create an immovable liquid boundary layer on the solid surface, in which the viscosity and the limiting shearing stress are much higher than those of the bulk fluids. While lower permeabilities mean shorter average pore radiuses, thinner pore throats may increase the proportion of the thickness of the boundary layer (the thickness of the hydrated layer for water) in the pore radius, thus reducing the flowing area in pores but increasing the percolating resistance for the driving fluids and at last, the TPG.

Figure 2-27 presents a linear relationship between TPG and permeability in log-log coordinates, showing that, although the intercepts on the vertical axis of different straight lines are different, their slopes are basically

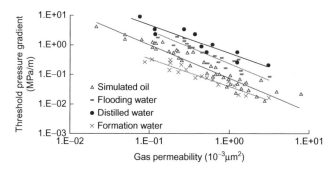

FIGURE 2-27   Relationship between TPG and permeability under different displacing fluids in a log-log plot.

consistent with each other, which is about -1, which can be proved by theoretical analysis.

The motion equation to express the influence of oil, water, and reservoir physicochemical properties on the seepage patterns is

$$v = \frac{10^{-6}K}{\mu}\left(1 - \frac{G}{\mathrm{grad}p}\right)\mathrm{grad}p \qquad (2.34)$$

At the same time, according to the physical significance of the TPG in ultralow-permeability reservoirs, the mathematical equation describing the flow characteristics can be expressed as

$$v = a \cdot \mathrm{grad}p - b.(\mathrm{grad}p \geq G) \qquad (2.35)$$

where

　　v—fluid velocity, m/s
　　μ—fluid viscosity, MPa·s
　　$\mathrm{grad}p$—pressure gradient, MPa/m
　　a and b—constant coefficients
　　There is a contrastive relationship between the two equations:

$$G = b\left(\frac{K}{\mu}\right)^{-1} \qquad (2.36)$$

After taking the logarithm of Eq. (2.36) we have

$$\log(G) = -\log(K) + \log(b \cdot \mu) \qquad (2.37)$$

Equation (2.37) indicates that the slope between the TPG line and the permeability line is -1, inversely proportional to each other, thus theoretically confirming the reliability of the data and the patterns obtained from the experiments.

2.2.3 An Analysis of Factors Affecting TPGs in Ultralow-Permeability Reservoirs

2.2.3.1 Differences in Fluid Types

In theory, under a fixed permeability, a higher fluid viscosity may involve a higher TPG. However, our experiments show that, under the same conditions, the lowest TPG is obtained from tests with formation water, the second lowest from model oil tests, the third lowest from injected water tests, and the highest from distilled water tests. At the same time, the permeabilities obtained from tests with model oil and formation water are roughly the same, higher than that from injected water and distilled water tests on the same scale. There may be two reasons for the fact that the TPGs of injected water and distilled water are much higher than those of model oil and formation water.

First, it may be related to clay expansion in the low-permeability core. The primary water from the formation has a salinity about 30 times of the injected water. The clay grains will expand, disperse, and escape when they encounter the less saline injected water, plugging the smaller pore throats, causing loss of their flow conductivity, or even completely destroying them. The permeability of the core therefore becomes poor while the filtrational resistance gets higher.

Second, it may be related to the clay solution caused by expansion and dispersion of clay grains. When the siliceous components enter into the water, the water close to the surface of the grain may become plastic fluid, whose viscosity will thus change. According to Einstein's viscosity law, the viscosity of the solution will increase significantly when the proportion of the clay volume in it increases to a certain point.

Therefore, the expansion, dispersion, and migration of the clay grains and the clay solution thus formed are a significant cause of nonlinear flow of injected water in the ultralow-permeability reservoir. We must apply anti-expansion treatment to the injected water at the injection wells.

2.2.3.2 The Influence of Reservoir Permeability and Oil Viscosity

Experiment results presented in Fig. 2-28 indicate that the TPGs with oils of different viscosities change in similar trends as the permeability varies but there are some differences. With a fixed permeability, a higher oil viscosity requires a higher TPG. At the same time, the differences between the TPGs obtained from different oils become smaller as the permeability increases.

The same permeabilities involve similar pore radiuses in the reservoir rocks. For the model oils made up of the same types of crude and kerosene, those with a higher viscosity may produce a thicker boundary layer. While oils with different viscosities have different structural mechanical properties, those with different structural mechanical properties have different limiting shearing stresses. In other words, the oil will not flow until its shearing stress

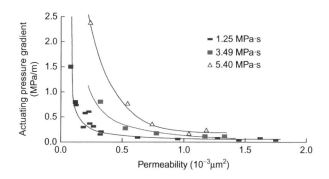

FIGURE 2-28   Different simulated oils with actuating pressure gradient.

is greater than the limiting shearing stress. Since one TPG can only drive the crude oil with a corresponding structural mechanics, and the limiting shearing stress of the high-viscosity oil is greater than that of the low-viscosity oil, it becomes natural that the former requires a greater pressure difference to be driven. This is why oil with a higher viscosity needs a higher TPG than one with a lower viscosity.

In the case that other conditions are kept the same, oil viscosity is proportional to TPG but inversely proportional to mobility changes caused by viscosity variations.

2.3 FEATURES OF NONLINEAR FLOWS IN ULTRALOW-PERMEABILITY RESERVOIRS

2.3.1 The Percolation Theory Involving Stress Sensitivity

2.3.1.1 Equations for Deliverability and Pressure Distribution of Steady Seepage Flows

For single-phase fluids, the equation for their areal radial flow velocities is

$$\frac{QB}{2\pi rh} = \frac{K}{\mu}\frac{dp}{dr} \tag{2.38}$$

Permeability is a function of effective overburden pressure with permeability variations considered. Putting Eq. (2.9) into Eq. (2.38):

$$\frac{Q_S B}{2\pi rh} = \frac{K_0}{\mu}\left(\frac{\sigma_v - \alpha p}{\sigma_{eff0}}\right)^{-S}\frac{dp}{dr} \tag{2.39}$$

The boundary conditions for steady seepage flows can be expressed as

$$\begin{cases} r = r_w, \ p = p_{wf} \\ r = r_e, \ p = p_e \end{cases} \tag{2.40}$$

Integrating the individual variables in Eq. (2.39) and putting the results into the formula for boundary conditions, we can get the formula for output:

$$Q_S = \frac{2\pi K_0 h}{\mu B \sigma_{eff0}^{-S}} \cdot \frac{(\sigma_v - \alpha p_{wf})^{1-S} - (\sigma_v - \alpha p_e)^{1-S}}{\alpha(1-S)\ln\frac{r_e}{r_w}} \tag{2.41}$$

We also get the formula for pressure distribution:

$$p(r) = \frac{\sigma_v - \left[\dfrac{Q\mu B\sigma_{eff0}^{-S}\alpha(1-S)\ln\dfrac{r_e}{r}}{2\pi K_0 h} + (\sigma_v - \alpha p_e)^{1-S}\right]^{\frac{1}{1-S}}}{\alpha} \tag{2.42}$$

where

Qs—flow rate, cm^3/s

B—volumetric factor

K_0—initial permeability, 10^{-3} μm^2

h—reservoir thickness, m

μ—fluid viscosity, mPa s

σ_{eff0}—initial effective stress (reference value), MPa

σ_v—overburden pressure, σv = ρrgH, MPa

pe—drainage boundary pressure, MPa

p_{wf}—bottom hole flowing pressure, MPa

$p(r)$—pressure at the point with a distance of r from the well hole, MPa

α—effective stress coefficient (ranges from 0.9—1.0 for low- and ultralow-permeability reservoirs)

S—stress-sensitivity coefficient

r_e and r_w—drainage radius and well radius, m

In fact, if we use the ground gas-logging permeability as the reference value for the initial permeability, then its variations caused by medium deformation may influence the deliverability in two ways. One is called the static influence, which refers to the influence of the overburden pressure on permeability variations, while the other is the dynamic influence, i.e., the influence of changes in the pore pressure during production on permeability variations, which attracts more attention from oil producers at present. Figures 2-29 to 2-32 show the curves of how stress sensitivity influences production.

2.3.1.2 Equations for Deliverability and Pressure Distribution of Pseudo-Steady Seepage Flows

Assume there is a well at the center of a closed-circle reservoir. Suppose the initial formation pressure in the drainage region is p_i and the average

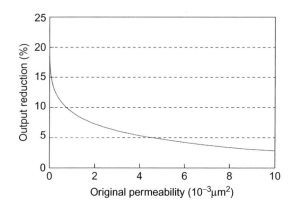

FIGURE 2-29   The relationship between the output decline percentage and initial permeability.

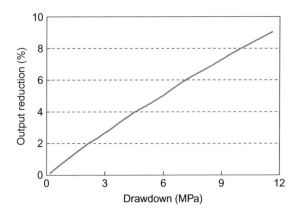

FIGURE 2-30   The relationship between output decline percentage and pressure drawdown.

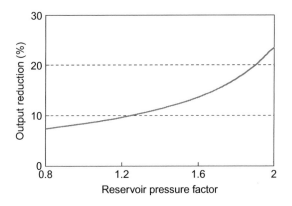

FIGURE 2-31   The relationship between output decline percentage and reservoir pressure factor.

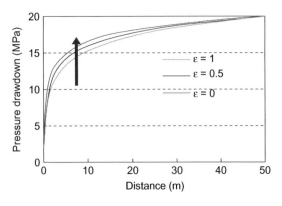

FIGURE 2-32   The pressure distribution curve considering stress sensitivity.

formation pressure is $\bar{p}_R$ after a period t of production. In a closed reservoir, the well output will completely depend on fluid expansion and pore volume reduction caused by pressure decline. According to the physical significance of the comprehensive compressibility C_t, the fluid volume driven out by the elastic energy of the drainage zone can be expressed as

$$V = C_t V_f (p_i - \bar{p}_R) \tag{2.43}$$

where

$V_f = \pi(r_e{}^2 - r_w{}^2)$—the rock volume in the drainage region

The well output is

$$Q = -\frac{dV}{dt} = -C_t \pi (r_e^2 - r_w^2) h \frac{d\bar{p}_R}{dt} \tag{2.44}$$

At the pseudo-steady stage the drawdown velocity $\frac{dp}{dt}$ at any point in the formation should be equivalent to each other, and then the flow rate through any cross-section with a radius of r is

$$q_r = -C_t \pi (r_e^2 - r^2) h \frac{d\bar{p}_R}{dt} \tag{2.45}$$

Combining Eq. (2.44) with (2.45):

$$\frac{q_r}{Q} = \frac{r_2^2 - r^2}{r_e^2 - r_w^2} \tag{2.46}$$

Since $r_w{}^2 \ll r_e{}^2$, then $r_e{}^2 - r_w{}^2 \approx r_e{}^2$, and this can be simplified as

$$q_r = \left(1 - \frac{r^2}{r_e^2}\right) Q \tag{2.47}$$

The flow rate at any point r on the cross-section is

$$v_r = \frac{q_r}{2\pi r h} = \frac{1}{2\pi r h}\left(1 - \frac{r^2}{r_e^2}\right) Q = \frac{Q}{2\pi r_e h}\left(\frac{r_e}{r} - \frac{r}{r_e}\right) \tag{2.48}$$

If the flow follows Darcy's Law, then

$$v_r = \frac{K}{\mu}\frac{dp}{dr} = \frac{Q}{2\pi r_e h}\left(\frac{r_e}{r} - \frac{r}{r_e}\right) \tag{2.49}$$

Considering that the permeability is a function of effective overburden pressure as the permeability varies, and taking into account the fluid volumetric factor, we put Eq. (2.9) into (2.49) for integrating the individual variables at intervals $r \to R_e$ and $p \to p_e(t)$.

$$\int_{p(r,t)}^{p_e(t)} \frac{K_0}{\mu B}\left(\frac{\sigma_v - \alpha p}{\sigma_{eff0}}\right)^{-S} dp = \frac{Q}{2\pi r_e h}\int_r^{r_e}\left(\frac{r_e}{r} - \frac{r}{r_e}\right) dr \tag{2.50}$$

The pressure at any point in the layer can be calculated as

$$p(r,t) = \frac{\sigma_v \left\{ \frac{Q\mu B \sigma_{eff0}^{-S} \alpha (1-S) \left[\ln \frac{r_e}{r} - \frac{1}{2} \left(1 - \frac{r^2}{r_e^2} \right) \right]}{2\pi K_0 h} + [\sigma_v - \alpha p_e(t)]^{1-S} \right\}^{\frac{1}{1-S}}}{\alpha} \tag{2.51}$$

When $r = r_w$, then $p(r,t) \to p_{wf}(t)$. As $r_w^2 \ll r_e^2$, with the item r_w^2/r_e^2 omitted, the bottom-hole pressure at any time t is

$$p_{wf}(t) = \frac{\sigma_v - \left\{ \frac{Q\mu B \sigma_{eff0}^{-S} \alpha (1-S) \left[\ln \frac{r_e}{r} - \frac{1}{2} \right]}{2\pi K_0 h} + [\sigma_v - \alpha p_e(t)]^{1-S} \right\}^{\frac{1}{1-S}}}{\alpha} \tag{2.52}$$

Then the output formula is

$$Q = \frac{2\pi K_0 h}{\mu B \sigma_{eff0}^{-S}} \cdot \frac{[\sigma_v - \alpha p_{wf}(t)]^{1-S} - [\sigma_v - \alpha p_e(t)]^{1-S}}{\alpha(1-s)\left(\ln \frac{r_e}{r_w} - \frac{1}{2} \right)} \tag{2.53}$$

where

p_i—initial formation pressure, MPa

$\bar{p}_R$—average reservoir pressure in the drainage zone after a period t of production, MPa

$p_e(t)$—boundary pressure at any time t, MPa

$p_{wf}(t)$—bottom-hole flowing pressure at any time t, MPa

$p(r,t)$—pressure at the point a distance r from the well at time t, MPa

The meanings of the other symbols are the same as those in (2.42).

2.3.2 The Percolation Theory Involving TPG of Ultralow-Permeability Reservoirs

2.3.2.1 Equations for Deliverability and Pressure Distribution of Steady Seepage Flows

Take the areal radial fluid flow, for example. Assume there is a complete production well at the center of a homogeneous, circular, and horizontal reservoir with a uniform thickness. Suppose the reservoir permeability is K, its thickness is h, its liquid viscosity is μ, its radius of circular drainage boundary is r_e, its drainage pressure is p_e, the radius of the well is r_w, and the bottom-hole flowing pressure is p_w.

To maintain the seepage continuity of the steady radial flow, the equation should look like this:

$$\frac{dv}{dr} + \frac{v}{r} = 0 \tag{2.54}$$

The equation for fluid motion is

$$v = \frac{K}{\mu}\left(\frac{dp}{dr} - G\right) \tag{2.55}$$

Combining the two equations above, we have

$$\frac{d^2p}{dr^2} + \frac{1}{r}\left(\frac{dp}{dr} - G\right) = 0 \tag{2.56}$$

Through substitution of variables, let $\psi = \dfrac{dp}{dr} - G$, and the general solution of this equation is

$$p = C_1 \ln r + Gr + C_2 \tag{2.57}$$

The boundary conditions can be derived from the physical model:

Inner boundary conditions $r = r_w$, $p = p_w$
Outer boundary conditions $r = r_e$, $p = p_e$

With the undetermined coefficients C_1 and C_2 determined by boundary conditions, the equation for pressure distribution of low-velocity non-Darcy steady flows in ultralow-permeability reservoirs looks like this:

$$\begin{aligned}
p &= p_w + \frac{(p_e + p_w) - G(r_e - r_w)}{\ln\dfrac{r_e}{r_w}} \cdot \ln\frac{r}{r_w} + G(r - r_w) \\
&= p_e - \frac{(p_e - p_w) - G(r_e - r_w)}{\ln\dfrac{r_e}{r_w}} \cdot \ln\frac{r_e}{r} - G(r - r_w)
\end{aligned} \tag{2.58}$$

And the equation for pressure gradient distribution is

$$\frac{dp}{dr} = \frac{(p_e - p_w) - G(r_e - r_w)}{\ln\dfrac{r_e}{r_w}} \cdot \frac{1}{r} + G \tag{2.59}$$

Figures 2-33 and 2-34 show contrastive curves of the formation pressure distribution and effective displacement gradient of a real well with Darcy flow and TPG considered, respectively. It can be seen that, with TPG considered, the formation pressure at places with the same radius drops down. Some of the energy lost in the flow process goes to overcome the additional resistance caused by the TPG. At the same time, the effective displacement

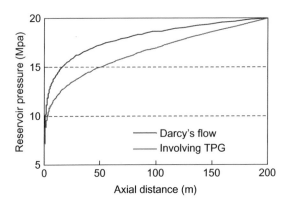

FIGURE 2-33　Comparison curves of the formation pressure distribution at a cross-section.

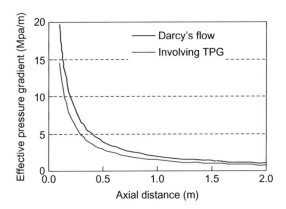

FIGURE 2-34　Effective displacement gradient distribution near the wellbore.

gradient of the whole reservoir is lower than that in the Darcy-flow case, with most pressure loss taking place near the wellbore.

Next is an equation for the areal radial fluid flow:

$$Q = 2\pi h v \tag{2.60}$$

Substituting equations for fluid motion and pressure gradient distribution into the equation above, we can obtain the formula of deliverability from the radial fluid flows in ultralow-permeability reservoirs:

$$Q = \frac{2\pi K h}{\mu} \cdot \frac{(p_e - p_w) - G(r_e - r_w)}{\ln \frac{r_e}{r_w}} \tag{2.61}$$

It can be seen from this equation that, in the case of stable seepage flows, the fluids have to overcome the additional resistance $G(r_e - r_w)$ caused by TPG and therefore may suffer additional energy losses, which will not contribute to well productivity. As a result, under the same circumstances, the well productivity involving TPG is lower than that from pure Darcy flows, and the higher the TPG is, the lower the well productivity will be.

2.3.2.2 Equations for Deliverability and Pressure Distribution of Unsteady Seepage Flows

Suppose that there is an oil well that produces at a fixed rate in the center of a homogeneous and infinite layer with uniform thickness, single-phased and compressible fluids, low-velocity non-Darcy Law seepage flows, the gravity and capillary forces neglected, and the temperature unchanged in the flowing process; then the seepage control equation, the initial conditions, and the boundary conditions can be, respectively, established as

$$\frac{\partial^2 p}{\partial r^2} + \frac{1}{r}\left(\frac{\partial p}{\partial r} - G\right) = \frac{1}{\eta}\frac{\partial p}{\partial t} \tag{2.62}$$

$$p(r,0) = p_e \tag{2.63}$$

$$\left(\frac{\partial p}{\partial r} - G\right)\Bigg|_{r=r_w} = \frac{Q\mu}{2\pi K h r_w} \tag{2.64}$$

$$\left(\frac{\partial p}{\partial r} - G\right)\Bigg|_{r=R(t)} = 0 \tag{2.65}$$

$$p = p_e \quad (r \geq R(t)) \tag{2.66}$$

Let $\psi = p - G(r - r_w)$; then Eq. (2.62) can be transformed to

$$\frac{\partial^2 \psi}{\partial r^2} + \frac{1}{r}\frac{\partial \psi}{\partial r} = \frac{1}{\eta}\frac{\partial \psi}{\partial t} \tag{2.67}$$

Introduce the intermediate function ξ, and let $\xi = \frac{r^2}{\eta t}$; then

$$\frac{\partial \psi}{\partial r} = \frac{24}{\eta t}\cdot\frac{\partial \psi}{\partial \xi} \tag{2.68a}$$

$$\frac{\partial^2 \psi}{\partial r^2} = \frac{2}{\eta t}\cdot\frac{\partial \psi}{\partial \xi} + \frac{4r^2}{(\eta t)^2}\cdot\frac{\partial^2 \psi}{\partial \xi^2} \tag{2.68b}$$

$$\frac{\partial \psi}{\partial t} = -\frac{r^2}{\eta t^2}\cdot\frac{\partial \psi}{\partial \xi} \tag{2.68c}$$

So Eq. (2.67) can be turned into

$$\frac{\partial^2 \psi}{\partial \xi^2} + \left(\frac{1}{4} + \frac{1}{\xi}\right)\frac{\partial \psi}{\partial \xi} = 0 \tag{2.69}$$

The solution of Eq. (2.69) is

$$\xi \frac{\partial \psi}{\partial \xi} = C_1 \cdot e^{-\frac{1}{4}\xi} \tag{2.70}$$

The undetermined coefficient C_1 will be determined next. Substitute the parameters ψ and ξ into Eq. (2.64):

$$\xi \frac{\partial \psi}{\partial \xi}\bigg|_{\xi=\frac{r_w^2}{\eta t}} = \frac{Q\mu B}{4\pi Kh} \tag{2.71}$$

Combine Eqs. (2.70) and (2.71), and the undetermined coefficient C_1 can be obtained:

$$C_1 = \frac{Q\mu B}{4\pi Kh} e^{\frac{r_w^2}{4\eta t}} \tag{2.72}$$

Substitute the equation above into Eq. (2.70); then

$$\frac{\partial \psi}{\partial \xi} = \frac{Q\mu B}{4\pi Kh} e^{\frac{r_w^2}{\eta t}} \cdot \frac{e^{-\frac{2}{4}\xi}}{\xi} \tag{2.73}$$

and when $r = r$, $\psi = p - G(r - r_w)$, $\xi = \frac{r^2}{\eta t}$; when $r = R(t)$, $\psi = p - G(R(t) - r_w)$, $\xi = \frac{R^2(t)}{\eta t}$.

Substitute them into Eq. (2-73) and integrate over the interval $[r, R(t)]$:

$$\int_{p-G(r-r_w)}^{p_e-G(R(t)-r_w)} d\psi = \int_{\frac{r^2}{\eta t}}^{\frac{R^2(t)}{\eta t}} \frac{Q\mu B}{4\pi Kh} e^{\frac{r_w^2}{4\eta t}} \cdot \frac{e^{-\frac{1}{4}\xi}}{\xi} d\xi \tag{2.74}$$

We notice that

$$\int_X^\infty \frac{e^{-u}}{u} du = -E_i(-x) \tag{2.75}$$

Finally, the pressure distribution at any time in the infinitely large reservoir can be obtained:

$$p = p_e - \frac{Q\mu B}{4\pi Kh} e^{\frac{r_w^2}{\eta t}} \left[-E_i\left(-\frac{r^2}{4\eta t}\right) + E_i\left(\frac{-R^2(t)}{4\eta t}\right) \right] - G[R(t) - r]$$

$$\approx p_e - \frac{Q\mu B}{4\pi Kh} \left[-E_i\left(-\frac{r^2}{4\eta t}\right) + E_i\left(-\frac{R^2(t)}{4\eta t}\right) \right] - G[R(t) - r] \tag{2.76}$$

Expanding Eq. $-E_i(-x)$ and substituting the analytic solution of p, the change of the moving boundary can be determined through the material balance equation:

$$Qt = \pi(R^2(t) - r_w^2)\phi h C_t \cdot \bar{Y} \tag{2.77}$$

where

$$\bar{Y} = \frac{Q\mu B}{2\pi KhR^2(t)}\int_{r_w}^{R(t)}\left[a_0 + a_1\frac{r^2}{4\eta t} + a_2\left(\frac{r^2}{4\eta t}\right)^2 + a_3\left(\frac{r^2}{4\eta t}\right)^3 + a_4\left(\frac{r^2}{4\eta t}\right)^4 + a_5\left(\frac{r^2}{4\eta t}\right)^5 + a_6\ln\left(\frac{r^2}{4\eta t}\right)\right]r\cdot dr$$

$$-\frac{Q\mu B}{2\pi Kh}\left[a_0' + a_1'\frac{R^2(t)}{4\eta t} + a_2'\left(\frac{R^2(t)}{4\eta t}\right)^2 + a_3'\left(\frac{R^2(t)}{4\eta t}\right)^3 + a_4'\left(\frac{R^2(t)}{4\eta t}\right)^4 + a_5'\left(\frac{R^2(t)}{4\eta t}\right)^5 + a_6'\ln\left(\frac{R^2(t)}{4\eta t}\right)\right]$$

$$+\frac{1}{3}R(t)G$$

and a_0 through a_6' are the coefficients related to $x = r^2/4\eta t$.

If the well produces at changing rates of $Q(t)$, in echelonment, then this kind of problem can be solved with the superposition principle.

According to the superposition principle, production at changing rates can be regarded as several wells in the same locality producing at the same time with different commissioning times. The first well produces with an output of $(Q_1-Q_0 = Q_1,(Q_0 = 0))$ from time $t_1 = 0$ to time t, accompanied by a pressure drop of Δp_1, which affects the outer boundary $R_1(t)$ of the active zone. The rest can be done in the same way. The well n produces from time t_n to time t with an output of (Q_n-Q_{n-1}), accompanied by a pressure drop of Δp_n, which affects the outer boundary $R_1(n)$ of the active zone. So the total pressure drop is

$$\Delta p = \Delta p_1 + \Delta p_2 + \cdots \Delta p_n$$

$$= \frac{(Q_1 - 0)\mu B}{4\pi Kh}e^{\frac{r_w^2}{4\eta(t-0)}}\left[-E_i\left(-\frac{r^2}{4\eta(t-0)}\right) + E_i\left(-\frac{R_1^2(t)}{4\eta(t-0)}\right)\right] + \lambda(R_1(t) - r_w + \cdots$$

$$+ \frac{(Q_n - Q_{n-1})\mu B}{4\mu Kh}e^{\frac{r_w^2}{4\eta(t-t_n)}}\left[-E_i\left(-\frac{r^2}{4\eta(t-t_n)}\right) + E_i\left(-\frac{R_n^2(t)}{4\eta(t-t_n)}\right)\right] + \lambda(R_n(t) - r_w)$$

$$= \frac{\mu B}{4\pi Kh}\left\{\sum_{i=1}^{n}\left[(Q_n - Q_{n=1})e^{\frac{r_w^2}{4\eta(t-t_i)}}\left[-E_i\left(-\frac{r^2}{4\eta(t-t_i)}\right) + E_i\left(-\frac{R_n^2(t)}{4\eta(t-t_i)}\right)\right] + \lambda(R_i(t) - r_w)\right]\right\}$$

$$\tag{2.78}$$

and the pressure at any time and any point of the layer is

$$p(r,t) = p_e - \Delta p$$

$$= p_e - \frac{\mu B}{4\pi Kh} \left\{ \sum_{i=1}^{n} \left[(Q_n - Q_{n-1})e^{\frac{r_w^2}{4\eta(t-t_i)}} \left[-E_i\left(-\frac{r^2}{4\eta(t-t_i)}\right) + E_i\left(-\frac{R_n^2(t)}{4(t-t_i)}\right) \right] - \lambda(R_i(t) - r_w) \right] \right\}$$

(2.79)

2.3.3 The Percolation Theory Involving Both TPG and Medium Deformation

2.3.3.1 Equations for Deliverability and Pressure Distribution of Steady Seepage Flows

The equation for fluid motion is

$$v = -\frac{K}{\mu}\left(\frac{dp}{dr} - G\right)$$

(2.80)

$$v = \frac{QB}{2\pi rh}$$

(2.81)

The laws of permeability variations along with changes in the formation pressure are

$$\frac{K}{K_0} = \left(\frac{\sigma_v - \alpha p}{\sigma_{eff0}}\right)^{-S}$$

(2.82)

With the help of the three equations above, we can obtain

$$\frac{QB}{2\pi rh} = -\frac{K_0}{\mu}\left(\frac{\sigma_v - \alpha p}{\sigma_{eff0}}\right)^{-S}\left(\frac{dp}{dr} - G\right)$$

(2.83)

This is an inhomogeneous equation with strong nonlinearity. Since a conventional analysis method fails to get an exact numerical solution, we can use an approximation method to solve it and obtain a solution for the steady seepage flows. Suppose that the pressure distribution can be expressed by the logarithm and exponential polynomial of coordinate r:

$$p = a_0 \ln\frac{r}{r_e} + a_1 + a_2\frac{r}{r_e}$$

(2.84)

Here are the boundary conditions:

$$p(r = r_w) = p_w$$

(2.85)

$$\left(\frac{\partial p}{\partial r} - G\right)\bigg|_{r=r_e} = \frac{Q\mu B}{2\pi K_0 \left(\frac{\sigma_v - \alpha p_e}{\sigma_{eff0}}\right)^{-S} h r_e} \tag{2.86}$$

$$p(r = r_e) = p_e \tag{2.87}$$

Substituting these conditions into the approximation equation for pressure distribution, we can get the deliverability equation for areal radial flows in the ultralow-permeability reservoir involving TPG and medium deformation:

$$Q = \frac{2\pi K_0 h (r_e - r_w)}{\alpha \mu B} \left[\frac{p_e - p_w - G(r_e - r_w)}{\dfrac{r_e - r_w + r_w \ln \dfrac{r_w}{r_e}}{\left(\dfrac{\sigma_v - \alpha p_e}{\sigma_{eff0}}\right)^{-S}} - \dfrac{r_e - r_w + r_e \ln \dfrac{r_w}{r_e}}{\left(\dfrac{p_c - \alpha p_w}{\sigma^*}\right)}} \right] \tag{2.88}$$

The formula for pressure distribution in the formation is

$$p = p_e - \frac{p_e - p_w - A\left(1 - \dfrac{r_w}{r_e}\right)}{B} \cdot \ln\frac{r}{r_e} - \frac{p_e - p_w + A\ln\dfrac{r_w}{r_e}}{B} \cdot \left(1 - \frac{r}{r_e}\right) \tag{2.89}$$

where $A = \dfrac{p_e - p_w - G(r_e - r_w)}{1 + \dfrac{r_w}{r_e - r_w}\ln\dfrac{r_w}{r_e} - \left(1 + \dfrac{r_e}{r_e - r_w}\ln\dfrac{r_w}{r_e}\right)\left(\dfrac{\sigma_v - \alpha p_e}{\sigma_v - \alpha p_w}\right)^{-S}} + Gr_e$

$$B = \ln\frac{r_w}{r_e} + 1 - \frac{r_w}{r_e}$$

The formula for pressure gradient distribution is

$$\frac{\partial p}{\partial r} = -\frac{p_e - p_w - A\left(1 - \dfrac{r_w}{r_e}\right)}{B} \cdot \frac{1}{r} + \frac{p_e - p_w + A\ln\dfrac{r_w}{r_e}}{B} \cdot \frac{1}{r_e} \tag{2.90}$$

With media deformation ignored, i.e., when $S = 0$, Eqs. (2.88) and (2.89) can be simplified into equations of deliverability and pressure distribution of steady single-phased seepage flows involving TPG only.

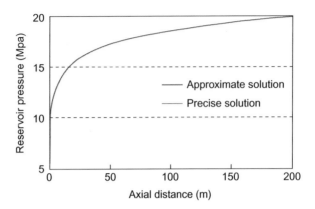

FIGURE 2-35 A comparison between approximate and exact solutions to formation pressure involving medium deformation.

If TPG is ignored, i.e., when $G = 0$, the approximate solution of the equation of steady seepage deliverability in deformable media can be obtained:

$$Q = \frac{2\pi K_0 h (r_e - r_w)}{\alpha \mu B} \left[\frac{p_e - p_w}{\dfrac{r_e - r_w + r_w \ln \dfrac{r_w}{r_e}}{\left(\dfrac{\sigma_v - \alpha p_e}{\sigma_{eff}0} \right)^{-S}} - \dfrac{r_e - r_w + r_e \ln \dfrac{r_w}{r_e}}{\left(\dfrac{\sigma_v - \alpha p_w}{\sigma_{eff}0} \right)^{-S}}} \right] \qquad (2.91)$$

Eq. (2.89) is still used to express the pressure distribution, except for

$$A = \frac{p_e - p_w}{1 + \dfrac{r_w}{r_e - r_w} \ln \dfrac{r_w}{r_e} - \left(1 + \dfrac{r_e}{r_e - r_w} \ln \dfrac{r_w}{r_e} \right) \left(\dfrac{\sigma_v - \alpha p_e}{\sigma_v - \alpha p_w} \right)^{-S}}$$

Figures 2-35 and 2-36 are the results of a comparison between the approximate and exact solutions to pressure distribution and deliverability involving medium deformation only, which suggest similar values of formation pressure distribution but slight differences in deliverability. The approximation method for pressure distribution and deliverability of ultralow-permeability reservoirs can solve the strong nonlinearity in the analytic method and achieve the precision required in production, but.

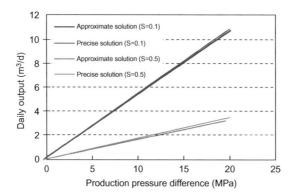

FIGURE 2-36   A comparison between approximate and exact solutions to productivity involving medium deformation.

2.3.3.2 Equations for Deliverability and Pressure Distribution of Unsteady Seepage Flows

Compared to medium- and high-permeability reservoirs, the most important feature of an ultralow-permeability reservoir is that the propagation of pressure waves in the layer lacks instantaneity, which means the transmission range of the formation energy is connected with the propagating time. To be more specific, the radius influenced by pressure drop of the production well is a function of time, i.e., the former increasing with the latter.

In this way, the whole reservoir can be divided into two sections in the radial direction at any time. i.e., the active zone within the pressure waves and the nonactive zone beyond the reach of the pressure waves. The outer boundary of the active zone can be seen as the drainage boundary at this moment. In the nonactive zone where the reservoir is not agitated, the pressure anywhere remains the same as what it is at the initial moment. When the boundary of the active zone reaches the natural boundary of the reservoir, the reservoir will produce steady or pseudo-steady seepage flows.

Assume that there is a producing well in the center of a homogeneous and infinitely large reservoir with a uniform thickness. If the changes in the viscosity of flows are ignored, then the pressure distribution in the active zone can be expressed with the logarithmic and exponential polynomial of coordinate r.

The equation for formation fluid continuity will look like this:

$$\frac{1}{r}\frac{\partial}{\partial r}(r\rho v) = -\frac{\partial}{\partial t}(\phi\rho) \tag{2.92}$$

with the stress-sensitivity coefficient of an ultralow-permeability reservoir included in the fluid motion equation:

$$v = -\frac{K_0}{\mu}\left[\frac{\sigma_v - \alpha p}{\sigma_{eff0}}\right]^{-S} \cdot \left(\frac{\partial p}{\partial r} - G\right) \tag{2.93}$$

Combining and expanding the two equations and ignoring the infinitely small items, we can get the equation for unstable flows in the ultralow-permeability reservoir involving TPG and medium deformation as

$$\left(\frac{\sigma_v - \alpha p}{\sigma_{eff0}}\right)^{-S} \cdot \left[\frac{\partial^2 p}{\partial r^2} + \frac{1}{r}\left(\frac{\partial p}{\partial r} - G\right)\right] = \frac{1}{\eta}\frac{\partial p}{\partial t} \tag{2.94}$$

where $\eta = (K_0/\phi\mu C_t)$.

Initial conditions:

$$p(r, 0) = p_e \tag{2.95}$$

Boundary conditions:

$$\left(\frac{\partial p}{\partial r} - G\right)\bigg|_{r=r_w} = \frac{Q(t)\mu B}{2\pi K_0\left(\dfrac{\sigma_v - \alpha p_w}{\sigma_{eff0}}\right)^{-S} hr_w}$$

$$\left(\frac{\partial p}{\partial r} - G\right)\bigg|_{r=R(t)} = 0 \tag{2.96}$$

$$p = p_e \quad (r \geq R(t))$$

According to the supposed conditions, the pressure distribution of the active zone can be expressed with the logarithmic and exponential polynomial of coordinate r, i.e.,

$$p = a_0 \ln\frac{r}{R(t)} + a_1 + a_2 \frac{r}{R(t)} + \cdots + a_{n+1}\frac{r^n}{R^n(t)}, \quad r_w \leq r \leq R(t) \tag{2.97}$$

According to the precision requirements here, we choose the first three items of n, i.e.,

$$p = a_0 \ln\frac{r}{R(t)} + a_1 + a_2 \frac{r}{R(t)}, \quad r_w \leq r \leq R(t) \tag{2.98}$$

Calculating the derivative of r in the equation above,

$$\frac{\partial p}{\partial r} = \frac{a_0}{r} + \frac{a_2}{R(t)} \tag{2.99}$$

According to the boundary conditions, we substitute Eqs. (2.98) and (2.99) into Eq. (2.96) to get three equations, which contain three unknown parameters, i.e., a_0, a_1, and a_2. Substituting them into the approximate equation of p, we get

$$p = p_e + \frac{Q(t)\mu B}{2\pi K_0 h}\left(\frac{\sigma_v - \alpha p_w}{\sigma_{eff0}}\right)^S \cdot \left(\ln\frac{r}{R(t)} + 1 - \frac{r}{R(t)}\right) - G(R(t) - r) \tag{2.100}$$

The key to the distribution of formation pressure is to obtain the movement pattern of $R(t)$, i.e., the outer boundary of the active zone.

When the Oil Well Produces at a Constant Rate

When the oil well produces at a constant rate, $Q(t) = Q = Const.$

Then the outer boundary of the active zone $R(t)$ can be obtained by the material balance equation, according to which the amount of fluid produced in a unit time is equal to changes of elastic fluid reserve in the active zone in the same period of time, i.e.,

$$Q = C_t \cdot \frac{d}{dt} \left[V(t) \overline{\Delta p} \right] \tag{2.101}$$

where

$$V(t) = \pi \left(R(t)^2 - r_w^2 \right) \phi h \tag{2.102}$$

$$\overline{\Delta p} = p_e - \overline{p} \tag{2.103}$$

The weighted average $\overline{p}$ in the active zone can be obtained by the equation as follows:

$$\overline{p} = \frac{1}{V(t)} \int_{V(t)} p(r,t) dV$$

$$= \frac{1}{\pi \left(R(t)^2 - r_w^2 \right) \phi h} \int_{r_w}^{R(t)} \left[p_e + \frac{Q\mu B}{2\pi K_0 h} \left(\frac{\sigma_v - \alpha p_w}{\sigma_{eff0}} \right)^{-S} \left(\ln \frac{r}{R(t)} + 1 - \frac{r}{R(t)} \right) - G(R(t) - r) \right] \cdot 2\pi \phi h r dr$$

$$= p_e - \frac{Q\mu B}{12\pi K_0 h} \left(\frac{\sigma_v - \alpha p_w}{\sigma_{eff0}} \right)^{-S} - \frac{1}{3} R(t) G$$

$$\tag{2.104}$$

Substitute Eqs. (2.102) to (2.104) into Eq. (2.101) and integrate it in the interval $[0, t]$:

$$12\eta t = \left(R(t)^2 - r_w^2 \right) \left[\left(\frac{\sigma_v - \alpha p_w}{\sigma_{eff0}} \right)^{-S} + \frac{4\pi K_0 h R(t) G}{Q\mu B} \right] \tag{2.105}$$

When the inner boundary condition is $r = r_w$, we substitute $p = p_w$ into the equation of p and

$$p_w = p_e + \frac{Q\mu B}{2\pi K_0 h} \left(\frac{\sigma_v - \alpha p_w}{\sigma_{eff0}} \right)^{-S} \cdot \left(\ln \frac{r_w}{R(t)} + 1 - \frac{r_w}{R(t)} \right) - G(R(t) - r_w) \tag{2.106}$$

Combining the two equations above containing the two unknown parameters p_w and $R(t)$, we can obtain the outer boundary radius $R(t)$ in the active zone at different times and then the pressure distribution in the formation at any time.

When the Oil Well Produces at Changing Rates

Integrating Eq. (2.93) in the interval $[r_w, R(t)]$ and substituting the boundary conditions (2-96) into it, we have

$$r\rho \frac{K_0}{\mu} \left(\frac{\sigma_v - \alpha p_w}{\sigma_{eff0}} \right)^{-S} \cdot \left(\frac{dp}{dr} - G \right) \bigg|_{r_w}^{R(t)} = \int_{r_w}^{R(t)} r\phi \frac{\partial \rho}{\partial t} dr \qquad (2.107)$$

$$\text{i.e.,} \quad -\frac{Q(t)B}{2\pi h \phi C_t} = \int_{r_w}^{R(t)} r \frac{\partial p}{\partial t} dr \qquad (2.108)$$

Integrating the equation above in the interval $[0, t]$ and substituting the boundary condition into it, we can obtain the rules of how the outer boundary radius $R(t)$ in the active zone changes over time under the condition of producing at changing rates:

$$\int_0^t Q(t) dt = (R(t)^2 - r_w^2) \left[\frac{Q(t)}{12\eta} \left(\frac{\sigma_v - \alpha p_w}{\sigma_{eff0}} \right)^{-S} + \frac{\pi \phi C_t h R(t) G}{3B} \right] \qquad (2.109)$$

In addition, according to the internal boundary conditions, we have

$$p_w = p_e + \frac{Q(t)\mu B}{2\pi K_0 h} \left(\frac{\sigma_v - \alpha p_w}{\sigma_{eff0}} \right)^{-S} \cdot \left(\ln \frac{r_w}{R(t)} + 1 - \frac{r_w}{R(t)} \right) - G(R(t) - r_w) \qquad (2.110)$$

Using a combination of Eqs. (2.109) and (2.110) we can solve the equation above. It can be seen that Eq. (2.105) is a special form of $Q(t) = C$ in Eq. (2.109).

When the Oil Well Produces under a Constant Bottom-Hole Pressure

Under this condition, the internal boundary conditions of unstable flows in ultralow-permeability reservoirs involving medium deformation and TPG are

$$p \big|_{r=r_w} = p_w, \qquad (2.111)$$

while other conditions are the same as those in Eqs. (2.93) to (2.95).

$$\text{Let } p = a_0 \ln \frac{r}{R(t)} + a_1 + a_2 \frac{r}{R(t)}, \qquad r_w \le r \le R(t) \qquad (2.112)$$

Substituting boundary conditions into Eq. (2.98) can determine the undetermined coefficients a_0, a_1, and a_2. Then

$$p = p_e - \frac{p_e - p_w - G(R(t) - r_w)}{\ln\dfrac{r_w}{R(t)} + 1 - \dfrac{r_w}{R(t)}} \ln\frac{r}{R(t)} - \frac{p_e - p_w + GR(t)\ln\dfrac{r_w}{R(t)}}{\ln\dfrac{r_w}{R(t)} + 1 - \dfrac{r_w}{R(t)}}\left(1 - \frac{r}{R(t)}\right)$$

(2.113)

Calculating the derivative of the equation above, we have

$$\frac{\partial p}{\partial r} = -\frac{p_e - p_w - G(R(t) - r_w)}{\ln\dfrac{r_w}{R(t)} + 1 - \dfrac{r_w}{R(t)}} \cdot \frac{1}{r} + \frac{p_e - p_w + GR(t)\ln\dfrac{r_w}{R(t)}}{\ln\dfrac{r_w}{R(t)} + 1 - \dfrac{r_w}{R(t)}} \cdot \frac{1}{R(t)}$$

(2.114)

Substituting this into the output equation, we have

$$Q(t) = 2\pi r_w h \cdot \frac{K_0}{\mu}\left(\frac{\sigma_v - \alpha p}{\sigma_{eff0}}\right)^{-S} \cdot \left(\frac{\partial p}{\partial r} - G\right)\bigg|_{r=r_w}$$

$$= 2\pi r_w h \cdot \frac{K_0}{\mu}\left(\frac{\sigma_v - \alpha p}{\sigma_{eff0}}\right)^{-S} \cdot \left[-\frac{p_e - p_w - G(R(t) - r_w)}{\ln\dfrac{r_w}{R(t)} + 1 - \dfrac{r_w}{R(t)}} \cdot \frac{1}{r_w} + \frac{p_e - p_w + GR(t)\ln\dfrac{r_w}{R(t)}}{\ln\dfrac{r_w}{R(t)} + 1 - \dfrac{r_w}{R(t)}} \cdot \frac{1}{R(t)} - G\right]$$

(2.115)

The equation for the change patterns of the outer boundary radius $R(t)$ over time in the active zone producing at changing rates can be transformed into

$$\int_0^{t-1} Q(t)dt + Q(t) \times 1 = (R(t)^2 - r_w^2)\left[\frac{Q(t)}{12\eta}\left(\frac{\sigma_v - \alpha p}{\sigma_{eff0}}\right)^{-S} + \frac{\pi\phi C_t h R(t) G}{3B}\right]$$

(2.116)

At time t, the output before time t-1 can be obtained through iteration, so the $\int_0^{t-1} Q(t)dt$ in the equation above is a known quantity.

Uniting Eqs. (2.115) and (2.116), which contain the two unknown parameters $Q(t)$ and $R(t)$, we can, through iteration and trial, get the boundary locality $R(t)$ of the active zone and the output at time t, $Q(t)$. Substituting the equation for $R(t)$ thus obtained into Eq. (2.113), we can at last obtain the pressure distribution in the formation.

2.4 APPLYING THE NONLINEAR PERCOLATION THEORY TO ADVANCED WATER INJECTION

Advanced injection means that the injection job begins before commissioning of oil wells. In other words, there is just waterflooding but no oil production during the lead time so that the formation pressure can be raised and a higher pressure gradient built when the oil wells are put into production. As a result, the pressure gradient at any point in the oil layer is higher than the TPG as the lead time reaches a certain value. The advantages of advanced water injection in reservoirs with low or ultralow permeabilities can be described as follows:

- It helps improve the two-phase seepage flows in the displacement of oil by water because it can raise the matrix permeability and activate the seepage flows in the microfracture system.
- It helps avoid irreversible damage caused by formation pressure drop.
- It helps reduce the seepage TPG and increase pressure drawdown to keep a high formation pressure so as to increase the maximum injector-producer distance and the effective swept region and establish an effective pressure driving system.
- As the seepage flows in low- or ultralow- permeability reservoirs have TPG, advanced injection helps overcome the viscous fingering during displacement of oil by water and improve the angle of the flow field, thus leading to even advancement of injected water and ultimately to a better result of oilfield development.
- It can prevent oil degassing caused by pressure drop and avoid the increase of seepage resistance.

2.4.1 The Influence of Advanced Injection on Permeability

2.4.1.1 The Influence on the Matrix Permeability

According to Eqs. (2.8) and (2.9), which are obtained from our experiments, we can calculate the influence of pressure maintenance levels on matrix permeability as shown in Fig. 2-37. Apparently, a lower level of pressure

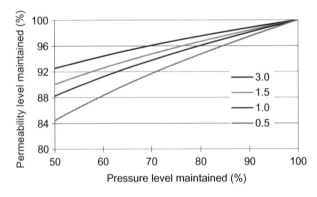

FIGURE 2-37 The relationship between levels of matrix permeability maintenance and pressure maintenance.

maintenance may result in a lower level of permeability maintenance. Moreover, at the same pressure maintenance level, a lower initial permeability will involve a lower level of permeability maintenance. Raising the pressure maintenance level through advanced injection may help maintain the permeability in ultralow-permeability reservoirs.

2.4.1.2 The Influence on Permeabilities of the Microfracture System

Most ultralow-permeability oilfields have developed fractures and are also called fractured ultralow-permeability oilfields. Fractures in the reservoirs have a double effect on waterflood production. On the one hand, they can improve the water-absorbing capacity of the injection well and the capacity of the production well. On the other hand, however, the fractures may cause fingering of the injected water along them, which will result in early water breakthroughs and explosive flooding in the production well.

Studies at home and abroad show that dramatic pressure drops in fractured ultralow-permeability reservoirs will cause porosity declines, fracture closings, and permeability reductions. Advanced injection, keeping the formation pressure at a higher level, helps reduce the damage to microfracture permeabilities caused by pressure drops. In addition, with a higher injection pressure differential, some of the closed microfractures will open again so that the permeability of the microfracture system may increase.

Observation of core samples shows that the reservoirs in the Yanchang Group contain some structural fractures, mainly with high angles and small in scale with lengths ranging from a few centimeters to dozens of centimeters. In addition, the group has quite developed interlayer fractures, some of which are open and may contribute to the improvement of horizontal permeabilities in the reservoir. As Table 2-2 shows, within the same rock sample, the section without fractures has a permeability of about $0.16 \times 10^{-3} \, \mu m^2$, but the section with developed fractures has a permeability up to $14.4 \times 10^{-3} \, \mu m^2$.

2.4.2 The Influence of Advanced Injection on TPG

TPG is influenced by the reservoir permeability, the pore-to-throat ratio, the fluid viscosity, and the saturation of a nonwetting phase. In a specific reservoir, advanced injection has an influence on the TPG in three aspects.

- Increase of the reservoir pressure may raise the matrix permeability (Fig. 2-38) but reduce its TPG. Besides, when the formation pressure exceeds the critical pressure for the microfracture to open, the permeability of the microfracture system may be improved to a larger degree while its TPG will drop dramatically.

TABLE 2-2 A Contrast of the Influence of Fractures on Porosities and Permeabilities in a Reservoir

Reservoir	Well no.	Serial number	Reservoir features	Porosity (%)	Permeability ($\times 10^{-3} \ \mu m^2$)
Chang-4 + 5	ZJ17	1	No fracture	11.05	0.16
		2		11.37	1.10
		3	Vertical fractures	11.60	14.40
	Yu-40-23	1	No fracture	11.31	1.42
		2	Horizontal fractures	11.71	4.24
Chang-8	X25	1	No fracture	7.90	0.66
		2	Fractures developed	12.10	5.75
	X24	1	No fracture	10.40	0.58
		2	Fractures developed	11.71	3.27

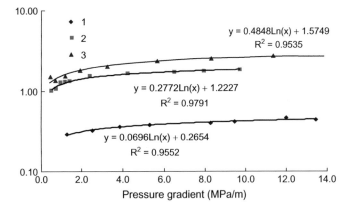

FIGURE 2-38 The relationship between pressure gradient and permeability.

- Advanced injection can raise the formation pressure and lower the oil viscosity, thus reducing the TPG and improving the mobility of oil in place.
- Advanced injection may greatly increase the water saturation around the water injector, thus changing the wettability of reservoir rocks and reducing the TPG.

TABLE 2-3 The Effect of Wettability on TPG

Serial number	Gas-logging permeability (10^{-3} μm^2)	Wettability	TPG (10^{-3} MPa/cm)
1	0.232	strongly water-wet	5.92
2		slightly water-wet	6.25
3		strongly water-wet	6.38
4	0.224	slightly oil-wet	6.70
5		strongly oil-wet	7.11

Note: The seepage fluid used for the experiment is the model oil with a viscosity of 1.25 mPa·s.

Practices show that different reservoir wettabilities and wetting degrees have a big influence on reservoir development. By measuring the TPGs of the cores with similar permeabilities but different wettabilities, we can study the influence of wettability on TPGs and then explain the influence of rock wettability on reservoir production and development from the prospective of TPG.

Through special treatment, the cores can become strongly water-wet, slightly water-wet, slightly oil-wet, and strongly oil-wet in turn. From Table 2-3 we can see that as the wettability of rock sample changes from water-wet to oil-wet, the TPG for the model oil to get through the core will also increase gradually, which is mainly influenced by the gravitational inter-action of the molecular interface. In the process from water-wet to oil-wet, the surface adsorptive force of the polar molecules in both the oil phase and the grains of porous media will increase. Under the same driving pressure gradient, the thickness of the layer of immovable fluids will also increase and the effective flow area in the pore throat will reduce while the flow resistance will rise, thus raising the TPG.

Chapter 3

Design of Advanced Water Injection in Ultralow-Permeability Reservoirs

Advanced Water Injection for Low Permeability Reservoirs.
© 2013 Petroleum Industry Press. Published by Elsevier Inc. All rights reserved.

The existence of the threshold pressure gradient (TPG) and medium deformation, which is the most prominent feature of ultralow-permeability reservoirs, will increase the filtrational resistance, reduce the per-well output, speed up the output decline, make it more difficult to stabilize the production, and thus lower the final recovery. On the basis of lab experiments, percolation theory, reservoir engineering, and field tests, this chapter will elaborate on the technological policies for advanced injection from such perspectives as well-pattern parameter optimization, injection timing, the commissioning time of producing wells, the maximum flowing pressure of water wells, the proper flowing pressure of oil wells, the timing for fracturing operations, and the requirements of advanced-injection practice.

3.1 THE WELL-PATTERN SYSTEM

3.1.1 Well Patterns

An areal well pattern is often adopted in the development of ultralow-permeability reservoirs due to their unique features, such as poor physical properties, low productivity, and the high driving pressure gradient needed for waterflooding. Therefore, a proper deployment of the injector-producer pattern serves as the foundation for efficient development. With the total well number fixed, different patterns will produce different effects.

Our practice with ultralow-permeability reservoirs of the Changqing Oilfield on the basis of in-depth studies of their geological and production characteristics not only successfully developed some large ultralow-permeability oilfields, such as Ansai, Jing'an, and Xifeng, but also put some immovable reserves into effective operation through constant optimization of well patterns and effective improvement of per-well output.

3.1.1.1 The History of Well-Pattern Deployment in Ultralow-Permeability Oilfields in Changqing

In our development of ultralow-permeability oilfields in Changqing, the optimal deployment of well patterns consisted of four main stages focused on matching well patterns and fractures; these patterns are discussed in the following.

The Square Inverted Nine-Spot Pattern with an Angle of 22.5° between the Well Array and the Fracture

The purpose of this well pattern is to slow down the water breakthrough and flooding along the fractures in oil wells. But because of the joint action of natural and artificial fractures, the injected water may break through along the fractures and form waterlines in the corner or edge producers adjacent to the water wells, which is hard to adjust. Wangyao, the first area that was put into production in the Ansai Oilfield, adopted this pattern with a well spacing ranging from 250 m to 300 m (Fig. 3-1a).

The Square Inverted Nine-Spot Pattern with the Well Array Parallel to the Fracture

Because the injector spacing in the main direction is the same as that on the lateral sides, the oil wells in the main direction showed a short response time,

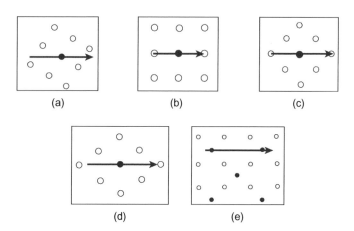

(a)   (b)   (c)

(d)   (e)

FIGURE 3-1 Evolution of well patterns in low-permeability oilfields in Changqing. (a) The square inverse nine-spot pattern with an angle of 22.5° between the well array and the fracture. (b) The square inverse nine-spot pattern with the well array parallel to the fracture. (c) The square inverse nine-spot pattern with an angle of 45° between the well array and the fractures. (d) The diamond pattern, with the longer diagonal of the rhombus parallel to the fracture direction. (e) The rectangular pattern, with the well array parallel to the fracture direction.

while those on the lateral sides made little response and showed a low-producing rate of reserves. The Pingqiao and Xinghe zones in Ansai adopted this pattern, with a well spacing ranging from 250 m to 300 m (Fig. 3-1b).

The Square Inverted Nine-Spot Well Pattern with an Angle of 45° between the Well Array and the Fractures

In this well pattern, the distance between oil producers and water injectors was extended in the main direction of the fractures to prolong the response time of the producers in this direction. But the producers on the lateral sides, on account of the large injector spacing, had even slower responses. Further, increasing the well spacing or narrowing the row spacing was restricted because the latter on the lateral sides had to be kept half of the former. This pattern was adopted in the Wuliwan zone in the Jing'an Oilfield, with a well spacing ranging from 300 m to 350 m (Fig. 3-1c).

The Diamond Inverted Nine-Spot and the Rectangular Well Patterns

In the diamond pattern (Fig. 3-1d), the longer diagonal of the rhombus is parallel to the direction of the fracture, which means the well spacing in the fracture direction is increased. Such deployment helps enlarge the fracturing scale, increase the length of artificial fractures, improve the per-well output, extend the stable production period, and slow down water out in corner producers. At the same time, it reduces the row spacing so that the oil wells on the lateral sides may have a better response. In the production tail, as the water-cut in the oil wells along the fracture rises to a certain level, the method of row-patterned drive is adopted to maximize the swept volume in matrix pores. This pattern produced good results in the Chang-6 reservoirs in the Panguliang zone of the Jing'an Oilfield and the Chang-8 reservoirs in the Baima zone of the Xifeng Oilfield.

But this pattern may cause quick water-out in the oil wells along the fractures. What is more serious is that certain wells in some blocks are drowned very soon after commissioning in the practice of a large-scale advanced injection. In view of this situation where oil wells on the fractures may be redundant and wasted, a rectangular well pattern (Fig. 3-1e) was proposed, in which the oil producers stood by but the fracturing scale was enlarged to practice linear injection. This pattern was tested in the ZJ60 zone in Wuliwan and produced good results.

3.1.1.2 Well Patterns for Reservoirs with Fractures That Are Not Well Developed

Three-square inverted nine-spot well patterns were adopted in Changqing:

• The pattern with an angle of 22.5° between the well array and the fracture (Fig. 3-1a)

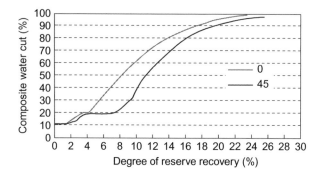

FIGURE 3-2 A comparison between the effects of different well arrays of square inverted nine-spot patterns.

- The pattern with the well array parallel to the fracture (with an angle of 0° between the well array and the fracture (Fig. 3-1b)
- The pattern with an angle of 45° between the well array and the fracture (Fig. 3-1c)

Numerical simulation through a geological model in view of the anisotropy of artificial fracture and permeability in the reservoirs shows that, as far as the square inverted nine-spot well patterns are concerned, the development indexes with an angle of 45° between the well array and the fracture are better than those with an angle of 0° (Fig. 3-2).

The areal square inverted nine-spot well pattern is adopted for blocks where natural microfractures are not very well developed and there is no obvious water breakthrough orientation, with the square diagonal parallel to the maximum stress direction (the angle between the well array and the fracture is 45°). In the Wuliwan area of the Jing'an Oilfield, all blocks, except for the test blocks of ZJ60, ZJ41, and ZJ53, which were developed through fracturing, used this pattern which, with a larger spacing between oil producers and water injectors in the main fracture orientation to slow down the water breakthrough in the direction, produced good results. But in this pattern, the well array spacing of oil producers on the lateral sides was so large that they may have relatively slow responses, and further increasing the well spacing or narrowing the row spacing was restricted because the latter on the lateral sides had to be kept half of the former.

3.1.1.3 Well Patterns for Reservoirs with Fractures That Are Well Developed

The areal injection pattern is often adopted to develop ultralow-permeability reservoirs, most of which, characterized by poor properties, well-developed natural fractures, significant permeability anisotropies, and low matrix

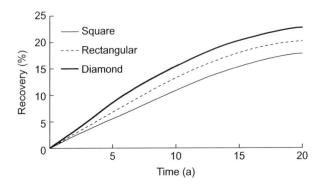

FIGURE 3-3 The recovery curves of different well patterns over time.

permeabilities, tend to show no productivity without artificial fracturing and need a great pressure gradient for development through water injection.

Considering the advantages and disadvantages of square inverted nine-spot well patterns as presented above, we conducted many lab experiments and adjustment tests to get a reasonable match between the well pattern and the fractures, which show that for ultralow-permeability zones with well-developed fractures, the line between the injector and the corner producer should be parallel to the fracture strike, with the well spacing in the fracture trend enlarged while the row spacing is narrowed down. So the diamond inverted nine-spot and rectangular patterns are proposed, which can help enlarge the fracturing scale, increase the length of artificial fractures, improve the per-well output and the stable production period, and prevent the corner producers from getting drowned too soon. At the same time, the two new patterns can help improve the response efficiency in the producers on the lateral sides so that a linear areal pattern can be adopted in the production tail to maximize the swept volume of matrix pores. Numerical simulation shows that, with the same density of the wells, the diamond inverted nine-spot pattern and rectangular pattern produce a higher producing rate and ultimate recovery than the square inverted nine-spot pattern (Fig. 3-3).

The Diamond Inverted Nine-Spot Well Pattern

The diamond pattern is adopted in blocks with well-developed fractures, with the longer diagonal of the diamond parallel to the direction of maximum principal stress so as to prevent the corner producers from getting flooded too soon and improve the response efficiency of the edge wells. At the same time, the enlarged well spacing in the fracture direction allows for a larger fracturing scale and longer artificial fractures, which helps increase the per-well output and the initial production. The features of the diamond pattern, such as high response efficiency, low water-cut, and higher oil recovery, make it more suitable for developing such reservoirs than the square pattern.

The Rectangular Well Pattern

The rectangular well pattern is adopted in the zones that contain well-developed fractures and show a clear direction of the maximum principal stress, with the well array parallel to fractures. Based on the inverted nine-spot pattern, the rectangular pattern, which can increase the per-well output by enlarging the fracturing scale, enhancing the injection intensity, and improving the flooding efficiency in the wells on the lateral sides of fractures, is an improved well pattern aimed at improving development of heterogeneous reservoirs.

With reference to different reservoir and fracture characteristics, the Changqing Oilfield adopts the square inverted nine-spot pattern, the diamond inverted nine-spot pattern, and the rectangular pattern, respectively, for reservoirs with undeveloped fractures, semi-developed fractures, and well-developed fractures so that the injector-producer patterns and the driving pressure system can well match the fracture system. In other words, fracturing, injection, and production are well integrated into these patterns.

3.1.1.4 Cases of Development

The Square Inverted Nine-Spot Pattern with the Diagonal Parallel to the Fracture Direction

This pattern is used in the whole Wuliwan area in the Jing'an Oilfield, except for the fracturing test wells of ZJ60, ZJ41, and ZJ53, with a well spacing of 330 m, the square diagonal parallel to the fracture, a sand volume of $25 \sim 40$ m^3 per well and an artificial fracture about 160 m long. The per-well output at different stages is shown in Table 3-1, with the water-drive-controlled reserves reaching 92%, the water-drive-produced reserves reaching 78%, a water-drive index of 1.62, a net injection percentage of 0.94, and a producer response rate of 81%. The formation pressure is maintained at 93% of its original, with a total water-cut of 18.6%, an oil-producing rate of 1.14%, a recovery ratio of 7.05%, and an expected recovery over 25%.

The Diamond Inverted Nine-Spot Well Pattern

The Wangyao area in Ansai. Under the same geological conditions and on the same fracturing scale, the Wang 25−05 well group, which adopted the 165×550 m diamond pattern, had a higher response rate, lower water-cut, and higher producing rate than its adjacent Wang 22-03 group in the west, which adopted the 300×300 m square pattern (Tables 3-2 and 3-3 and Fig. 3-4).

The Xinghe area in Ansai. In its early development the square inverted nine-spot pattern was adopted in this area, at an angle of 45° between the diagonal and the fracture. Later, when the Lian-12 well block was put into production, the diamond inverted nine-spot well pattern, with a well spacing

TABLE 3-1 Results of Development with Square Inverted Nine-Spot Pattern in Chang-6 Reservoirs of the Wuliwan Area

Well-testing stage		Initial stage			Half a year later			One year later			December 2007		
Oil (t/d)	Water (m³/d)	Fluid output (m³/d)	Oil output (t/d)	Water-cut (%)	Fluid output (m³/d)	Oil output (t/d)	Water-cut (%)	Fluid output (m³/d)	Oil output (t/d)	Water-cut (%)	Fluid output (m³/d)	Oil output (t/d)	Water-cut (%)
18.5	0.3	8.2	6.6	4.5	6.8	5.5	3.5	5.9	4.8	3.2	6.2	4.3	

TABLE 3-2 A Comparison of the Results of Development in Wang 25−05 and Wang 22-03 Well Groups (Table I)

Well group	Well pattern	Commissioning time	Response rate (%)	Water-cut (%)		Oil-producing rate (%)
				Initial stage	December 2007	Initial stage
Wang 25−05	Diamond inverted nine-spot	December 1996	100	5.1	11.1	1.59
Wang 22−03	Square inverted nine-spot	November 1996	87.5	16.1	34.2	1.05

TABLE 3-3 A Comparison of the Results of Development in Wang 25−05 and Wang 22−03 Well Groups (Table II)

Well pattern	Well-testing (fluid drainage) stage		Initial stage		Half a year later		One year later		December 2007		Cumulative oil output (10^4 t)
	Oil output (t/d)	Water output (m³/d)	Oil output (t/d)	Water-cut (%)	Oil output (t/d)	Water-cut (%)	Oil output (t/d)	Water-cut (%)	Oil output (t/d)	Water-cut (%)	
Square	17.46	2.02	3.1	18.45	2.88	14.8	2.6	13.96	1.875	46.45	0.4294
Diamond	16.117	2.01	3.51	6.75	2.86	5.1	2.41	4.55	2.165	35.05	0.6886

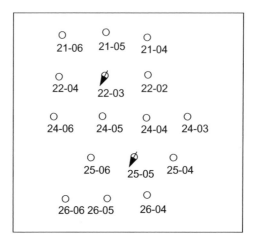

FIGURE 3-4　Well locations of the Wang 25-05 and Wang 22-03 well groups.

of 500 m and a row spacing of 150 m, was adopted due to the relatively poor reservoir properties in the zone and the initial success of this pattern in other blocks. A comparison between the new block of Lian-12 and the old block of Sai-126, whose permeabilities and reservoir thickness are similar, shows that the diamond pattern has an obvious advantage over the square pattern in that the former yields a higher per-well output and a lower decline rate (Table 3-4). To be more specific, the average single-well output in the diamond pattern is 0.36 t/d higher than that in the square pattern at the initial stage, 1.25 t/d higher six months later, and 1.38 t/d higher one year later. The decline rate in the diamond pattern is 13.5% one year later, as opposed to 36.3% in the square pattern.

The Rectangular Well Pattern

The Wuliwan area in Jing'an. The ZJ60 well block in this area adopted the rectangular well pattern with an oil-producer spacing of 480 m, a water-injector spacing of 960 m and a row spacing of 165 m, with the well array parallel to the maximum principal stress direction (NE 70°). Fracturing engineering was carried out for both the water injectors and oil producers, and the single-well sand input increased from 20 m^3 to 40 m^3 and the artificial fracture from 100 m to 160 m.

After waterflooding began in the test area in October 1998, the oil producers began to respond in February 1999 successively. By the end of June 2001, 40 producing wells had been flooded to different degrees, accounting for 95.2% of all oil producers in the test area, 10.4% higher than those in adjacent blocks.

TABLE 3-4 A Comparison of the Results of Development in the Xinghe Area in Ansai with Different Well Patterns

Well block	Well-testing (fluid drainage) stage		Initial stage		Six months later		One year later		December 2007		Cumulative oil output (10^4t)
	Oil output (t/d)	Water output (m^3/d)	Oil output (t/d)	Water-cut (%)	Oil output (t/d)	Water-cut (%)	Oil output (t/d)	Water-cut (%)	Oil output (t/d)	Water-cut (%)	
Lian-12 (diamond 500 × 150)	35.41	1.93	5.04	13.27	4.96	12.14	4.36	11.95	4.65	16.83	0.3238
Sai-126 (square 300 m)	12.49	1.36	4.68	3.06	3.71	2.73	2.98	2.34	2.52	25.36	0.8633

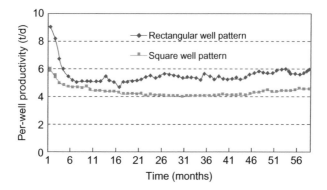

FIGURE 3-5 A comparison of per-well outputs with different well patterns in the Wuliwan area.

The change pattern of per-well output shows a better result of development in the test area in that its output at the initial stage is about 1.0 t/d higher than that in the neighboring block and 2.0 t/d higher after two years of production until 100 months (Fig. 3-5 and Table 3-5), which suggests that the rectangular well pattern, with a higher response rate and a longer period of stable production, is more suitable for such reservoirs and is a good prospect for dissemination.

The Sai-130 well block of Wangyao in Ansai. The Chang-6 reservoir in this area has an average effective thickness of 18.1 m, with a core permeability of only $0.37 \times 10^{-3} \, \mu m^2$. With reference to its poor reservoir properties and well-developed natural microfractures, we adopted the rectangular well pattern, with an injector-producer spacing of 1040 m and a row spacing of 120 m. Both the water injectors and oil producers were put into operation after sand fracturing. The water injector received an average sand input of 10.4 m³, at a sand-to-fluid ratio of 30.4% and a delivery rate of 1.6 m³/min, and the oil producer received an average sand input of 44.6 m³, at a sand-to-fluid ratio of 36.7% and a delivery rate of 2.4 m³/min. The average per-well output was 3.46 t/d at the initial stage and dropped to 2.46 t/d one year later (Table 3-6), which shows a good result of development.

The Nanliang Oilfield. The Chang-4 + 5 reservoir in the oilfield has poor reservoir properties and a core permeability of $0.49 \times 10^{-3} \, \mu m^2$. At the initial stage of development we used the rectangular well pattern, in which water breakthrough was seen in some oil producers in the fracture trend. This implies that the reservoir contains well-developed fractures. So later we shifted to the diamond well pattern, which soon increased the per-well output while lowering the water-cut (Table 3-7). Practice has shown that the diamond pattern is more suitable for this reservoir.

TABLE 3-5 A Comparison of the Results of Development with Different Well Patterns in Chang-6 Reservoirs of the Wuliwan Area

Well pattern	Core Permeability (10^{-3} μm^2)	Effective thickness (m)	Well-testing stage		Initial stage		Six months later		One year later		December 2007	
			Oil output (t/d)	Water output (m³/d)	Oil output (t/d)	Water-cut (%)	Oil output (t/d)	Water-cut (%)	Oil output (t/d)	Water-cut (%)	Oil output (t/d)	Water-cut (%)
Rectangular 960 × 360 m	1.4	12.1	21.2	0.1	7.6	4.9	6.4	4.8	5.7	4.1	6.2	7.4
Square 330 m	1.8	13.1	18.5	0.3	6.6	4.5	5.5	3.5	4.8	3.2	4.3	18.0

TABLE 3-6 Results of Development in the Sai-13 Block in Wangyao of Ansai with the Rectangular Well Pattern

Core analysis		Effective thickness		Well-testing stage		Initial stage		Six months later		One year later		December 2007	
Porosity (%)	Permeability (10^{-3} μm^2)	Oil layer (m)	Oil-water layer (m)	Oil output (t/d)	Water output (m^3/d)	Oil output (t/d)	Water-cut (%)	Oil output (t/d)	Water-cut (%)	Oil output (t/d)	Water-cut (%)	Oil output (t/d)	Water-cut (%)
10.56	0.37	10.5	7.6	19.1	4.8	3.46	22.7	2.82	21.3	2.4	19.3	2.04	14.6

TABLE 3-7 A Comparison of the Results of Development with Different Well Patterns in Nanliang Oilfield

Well pattern	Total of wells	Well-testing stage		Initial stage		Six months later		One year later		December 2007	
		Oil output (t/d)	Water output (m^3/d)	Oil output (t/d)	Water-cut (%)	Oil output (t/d)	Water-cut (%)	Oil output (t/d)	Water-cut (%)	Oil output (t/d)	Water-cut (%)
Diamond	14	19.2	0.3	3.8	21.3	3.1	15.4	3.3	17.0	3.1	21.4
Rectangular	21	22.9	1.4	4.4	13.1	3.4	11.5	3.7	12.0	4.5	12.5

3.1.2 Reasonable Row Spacing

Reasonable row spacing in a well pattern helps build a reasonable injection and production pressure difference so as to achieve better injection results. The row spacing for the development of ultralow-permeability reservoirs mainly depends on the permeability of their matrix rocks and the fracture density. Lower permeabilities and smaller densities require narrower row spacing, and vice versa.

3.1.2.1 Theoretical Calculation of Reasonable Injector-Producer Spacing

Reasonable well spacing depends on four factors: reservoir permeability, reservoir fluid viscosity, injection and production pressure difference, and expected oil output. While the pressure gradient is determined by the first two factors, the permeability and effective overburden pressure affect the stress sensitivity.

According to the oil productivity equation (2.88), under certain reservoir conditions (i.e., a fixed permeability, fluid viscosity, and production pressure difference, which determine the TPG and stress-sensitivity factor) and stable production conditions, the drainage radius is closely related to a unique liquid output. In other words, there must be a maximum drainage radius to satisfy a minimum liquid producing rate.

Take the square five-point well pattern as an example. Under the conditions of a stable oil production and a balance between injection and production, the injector-producer distance may be regarded as equivalent to the drainage radius if the well groups are symmetrical. This can then be converted to the maximum fluid supply in the research of a balanced off-take between one injector and one producer (Fig. 3-6).

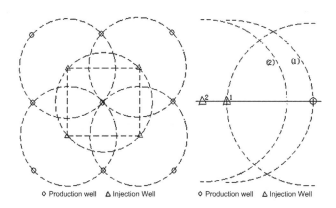

◇ Production well   △ Injection Well   ◇ Production well   △ Injection Well

FIGURE 3-6   Diagram of flow boundary in a simplified well pattern.

According to the superposition theory, the pressure at any point within the swept region of one injector and one producer is expressed as

$$p = p_e - \frac{p_e - p_w - A\left(1 - \dfrac{r_w}{r_{ew}}\right)}{B} \cdot \ln\frac{\sqrt{(x-x_w)^2 + (y-y_w)^2}}{r_{ew}} - \frac{p_e - p_w + A\ln\dfrac{r_w}{r_{ew}}}{B} \cdot \left(1 - \frac{r}{r_{ew}}\right)$$

$$+ \frac{p_h - p_e - A\left(1 - \dfrac{r_w}{r_{eh}}\right)}{B} \cdot \ln\frac{\sqrt{(x-x_h)^2 + (y-y_h)^2}}{r_{eh}} + \frac{p_e - p_w + A\ln\dfrac{r_w}{r_{eh}}}{B} \cdot \left(1 - \frac{r}{r_{eh}}\right)$$

(3.1)

where

$$A = \frac{p_e - p_w - G(r_e - r_w)}{1 + \dfrac{r_w}{r_e - r_w}\ln\dfrac{r_w}{r_e} - \left(1 + \dfrac{r_e}{r_e - r_w}\ln\dfrac{r_w}{r_e}\right)\left(\dfrac{\sigma_v - \alpha p_e}{\sigma_v - \alpha p_w}\right)^{-S}} + Gr_e$$

$$B = \ln\frac{r_w}{r_e} + 1 - \frac{r_w}{r_e}$$

Thus the pressure gradient on the borehole face of the producer and its output can be obtained.

If an injector and a producer are beyond the pressure swept region of the other, which means the well drainage radius exceeds the maximum distance to stabilize the liquid feed for the current output, the oil output will decline, to strike a new balance.

In this way a reasonable injector-producer distance can be determined in the cases of different reservoir permeabilities, fluid viscosities, injection and production pressures, and expected oil outputs.

Reservoir Parameters

The original formation pressure 20 MPa, the injection and production pressure difference 20 MPa, the oil viscosity 1 mPa · s, the density 0.85 g/cm3 and the volume factor 1.2. The oil TPG and the stress-sensitivity coefficient have been determined by the experiments in Chapters 2.

One single map alone cannot give an adequate explanation for how the injector-producer spacing is influenced by four factors, i.e., reservoir permeability, fluid viscosity, injection and production pressure, and expected output. Fig. 3-7 presents the relations between reasonable well spacing on the one hand and the permeability levels and desired outputs on the other under the condition of a given fluid viscosity and injection and production pressure. It can be seen that with a given permeability, the higher the expected output, the smaller the reasonable well spacing, and with a given expected output, a

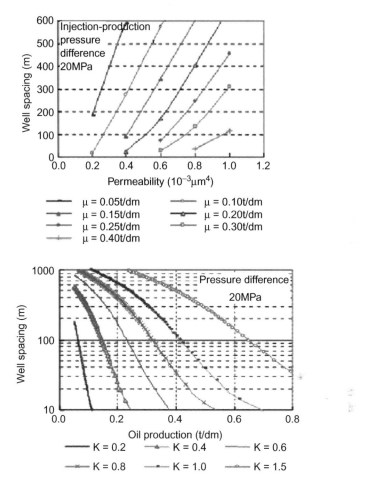

FIGURE 3-7 Maps of reasonable injector-producer spacing to match different permeabilities and expected outputs.

greater permeability means that the injector-producer spacing can be increased accordingly.

3.1.2.2 Reasonable Row Spacing for an Effective Pressure Displacement System

For ultralow-permeability reservoirs with developed fractures, the key to an effective pressure displacement system is to solve the problem of lateral oil displacement, which requires reasonable row spacing.

A large number of laboratory tests show that oil and gas flows in an ultra-low permeability reservoir have a non-Darcy flow with a TPG. According to the experiment data, a lower permeability entails a greater TPG, which will

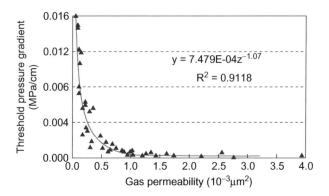

FIGURE 3-8   Curve of the relationship between the permeability and the TPG.

increase sharply when the permeability drops to below $0.5 \times 10^{-3}\,\mu m^2$ (Fig. 3-8). So determining reasonable row spacing according to the TPG is the prerequisite for an effective pressure displacement system. The lower the permeability, the smaller the row spacing should be.

Figure 3-9 presents the curves obtained through theoretical formulas to describe the pressure distribution between the injector and the producer with different row spacing for reservoirs of different permeabilities and their corresponding formation pressure gradients. It can be seen that the pressure gradient at any place in the reservoir should be greater than the TPG to establish an effective pressure driving system.

3.1.2.3 Reservoir Engineering Calculations for Reasonable Injector-Producer Spacing

The Relationship between Well Density and Recovery Factor

Researchers such as PENG Changshui and GAO Wenjun (2000) established an oil recovery equation involving the swept volume coefficient, well density, injector-producer ratio, well pattern parameters, and reservoir properties, thus providing a relatively comprehensive solution to show the effects of these factors on recovery, which is expressed as

$$E_V = e^{\dfrac{-b(\phi S_o h S_c)^x}{R^{0.5x} K_e^{1.5x} A_c^z}}, \quad a = 2.16894 \times b \tag{3.2}$$

Mathematical statistics and theoretical studies show that the swept volume coefficient and the well density form a linear equation, i.e., $x = 1$, and according to Fan Jiang $et\ al.$, $z = 1$.

Therefore, Eq. (3.2) can be changed to

$$E_V = e^{\dfrac{-a\phi S_o h S_c}{R^{0.5} K_e^{1.5} A_c}} \tag{3.3}$$

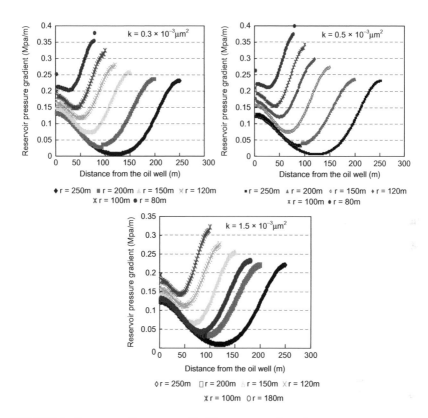

FIGURE 3-9 Pressure gradient curves with different row spacing schemes.

This equation encompasses all the domestic research results available, including the waterflood swept volume coefficient, well density, injector-producer ratio, well-pattern parameters, and reservoir properties, thus providing a much more comprehensive way to present the impact of different factors on the recovery. Usually the recovery factor can be expressed by the product of the oil displacement efficiency and the swept volume coefficient, but as the lithological reservoir at issue has poor continuity in the lateral direction and small oil-bearing sandbodies, some Chinese researchers have expressed the recovery factor in ultralow-permeability reservoirs as follows:

$$E_R = E_c \cdot E_d \cdot E_v$$

$$E_R = E_c \cdot E_d e^{\dfrac{-a\phi S_o h S_c}{R^{0.5} K_e^{1.5} A_c}} \tag{3.4}$$

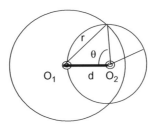

FIGURE 3-10   The relationship between the reservoir continuity and the reserves radius.

where

> E_d—oil displacement efficiency, f
> ϕ—porosity, f
> a—coefficient to be determined, dimensionless, usually between 2 and 3
> K_e—effective permeability of the oil layer, $10^{-3}\ \mu m^2$
> S_o—oil saturation, f
> h—effective thickness of the oil layer, m
> A_c—well-pattern parameters, f
> S_c—well density, km^2/well
> R—injector-producer ratio, f

E_c, which means reservoir continuity, can be calculated this way: in the case of using a square well pattern with a well spacing of d for the development of a reservoir with reserves radius r (Fig. 3-10), we can form the expression

$$E_c = \frac{2}{\pi} \arccos \frac{d}{2r} - \frac{1}{2\pi r^2} \sqrt{4d^2 r^2 - d^4} \qquad (3.5)$$

According to this equation, as far as an injector-producer spacing of 500 m is concerned, a reserves radius from 2000 m to 3000 m means a reservoir continuity between 85% and 90%. Since the reserves radius in the research area usually remains less than 3000 m, the reservoir continuity is often less than 90%. But if the reserves radius reaches 4000 m, the reservoir continuity can increase up to 95% even if the injector-producer spacing is kept as small as 300 m. Thus a well pattern should fit the spatial distribution of oil layers for the development of ultralow-permeability lithological reservoirs.

Reasonable Well Density Research

Although a higher well density may bring about a higher recovery, the increase of recovery will slow down and therefore become economically unworthy after the well density reaches a certain value. This value is then called the reasonable, or minimum, well density.

Suppose the total costs of drilling and ground construction for a single well are F yuan; then the total investment for a whole oilfield is $A \times f \times F$.

Suppose the crude production costs C yuan/t; then the costs of oil production total $N \times E_R \times C$. If the after-tax price of crude oil is P yuan/t, then the sales income from the developed field is expressed as $N \times E_R \times P$. With an interest rate of R for investment loans and an evaluation period of T, the total profit for the oilfield development can be expressed as

$$M = N \cdot E_R P - A \cdot f \cdot F - N \cdot E_R \cdot C(1+R)^{T/2} \tag{3.6}$$

Putting Eq. (3.4) into (3.6) we get

$$M = N \cdot E_R(P - C) - A \cdot f \cdot F(1+R)^{T/2} \tag{3.7}$$

For the optimal well density, derivate f in Eq. (3.7) to get

$$\frac{\partial M}{\partial f} = \frac{N \cdot I_X \cdot E_R}{f^2}(P - C) - A \cdot F(1+R)^{T/2} \tag{3.8}$$

Therefore, an optimal well-spacing density should satisfy the following transcendental equation, which needs to be solved through iteration:

$$\frac{N \cdot I_X \cdot E_R}{f_{opt}^{~2}}(P - C) - A \cdot F(1+R)^{T/2} = 0 \tag{3.9}$$

in which

$$I_X = \frac{-a\phi S_o h}{R^{0.5} K_e^{1.5} A_c}$$

By classifying the ultralow-permeability reservoirs of Chang-8, Chang-6, and Chang 4 +5 in the six areas, i.e., middle Baima, Dalugou-2, central and southern Xinghe, and Baiyushan (Table 3-8), and putting their reservoir parameters into Eq. (3.9), we can get the optimal well density at different prices of crude oil (Fig. 3-11). It can be seen that, as the oil prices go up, the optimal well density will increase, with the corresponding reasonable row spacing narrowing down.

Reasonable Row Spacing

After obtaining the optimal well density, we can calculate the optimal controllable area of a well, which is expressed as

$$A_{opt} = \frac{1}{f_{opt}} = d_x \cdot d_y \tag{3.10}$$

If the fracturing of the oil producer and the water injector are not considered, the well spacing and row spacing in an anisotropic formation should be adjusted as

$$\frac{d_x/2}{d_y} = \sqrt{\frac{K_x}{K_y}} \tag{3.11}$$

TABLE 3-8 Geological Properties of Four Types of Reservoirs

Reservoir type	Effective thickness (m)	Porosity (%)	Permeability ($\times 10^{-3}$ μm²)	Oil saturation (%)	Abundance ($\times 10^4$ t/km²)
Type-I	13.02~18.66	>11	>1.2	51.17~70.48	59.59~100.1
Type-II	11.71~18.09	10.1~12.85	1~1.7	50.64~67.35	56.23~90.46
Type-III	9.69~16.52	9.54~12.08	0.7~1.5	50.08~64.17	45.5~68.34
Type-IV	5.26~14.19	<10.87	<1	48.34~60.63	39.07~60.99

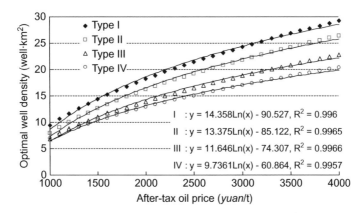

FIGURE 3-11   The relationship between crude prices after-tax and well density.

in which K_x and K_y are the permeability in the directions of x and y, respectively, $\times 10^3 \ \mu m^2$, and d_x and d_y are the well spacing (row spacing) in the directions of x and y respectively, m.

Considering fracturing and well-pattern factors, we can, by numerical simulation, get the relations between the anisotropy of the diamond inverted nine-spot pattern and interlaced-row pattern on the one hand and the reasonable well-spacing (d_x)/row-spacing (d_y) ratio (d_x/d_y) on the other (Fig. 3-11). Figure 3-11 shows that a higher reservoir permeability requires a lower well-spacing/row-spacing ratio under the conditions of the same reservoir matrix anisotropy and artificial fracturing scale. This is because with a given fracture density (namely, the fracture parameters), the lower the matrix permeability, the bigger the anisotropy of reservoir permeability.

We can now use the fitting equation in Fig. 3-12 to calculate the well-spacing/row-spacing ratio for different types of reservoirs with different permeabilities and different anisotropies (Table 3-9), and then, according to the relations between this ratio and the well density, obtain the reasonable row spacing for reservoirs with different anisotropies and different well patterns (Figs. 3.13 to 3.16).

Equations (3.10) and (3.11) can help calculate the well spacing and row spacing under the condition of different crude oil prices, different optimal well densities, and different anisotropies of the reservoir.

To sum up, we use the methods of reservoir engineering and numerical simulation to get the relationship between the recovery and the well density and the input-output theory to get the optimal well density at different crude prices. At the same time, the relationship between reservoir anisotropy and reasonable well and row spacing is established, and reasonable well and row spacing in different reservoir anisotropies is obtained through the relationship of well and row spacing to well density. In this way we establish a criterion

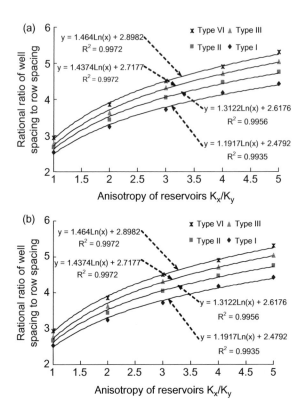

FIGURE 3-12 Relations between the anisotropy of reservoirs with different permeabilities (producer fracturing considered) and the reasonable row spacing. (a) Diamond inverted nine-spot well pattern. (b) Interlaced-row well pattern.

TABLE 3-9 A Reference Table of Row Spacing for Reservoirs with Different Permeabilities and Anisotropies

Well-spacing/row-spacing ratio $d_x:d_y$	Diamond inverted nine-spot pattern			Interlaced-row pattern		
	$K_x = 5K_y$	$K_x = 3K_y$	$K_x = 1K_y$	$K_x = 5K_y$	$K_x = 3K_y$	$K_x = 1K_y$
Type-I	4.40	3.79	2.48	3.77	3.34	2.41
Type-II	4.73	4.06	2.62	4.16	3.63	2.50
Type-III	5.03	4.30	2.72	4.73	4.07	2.65
Type-IV	5.25	4.51	2.90	5.08	4.37	2.83

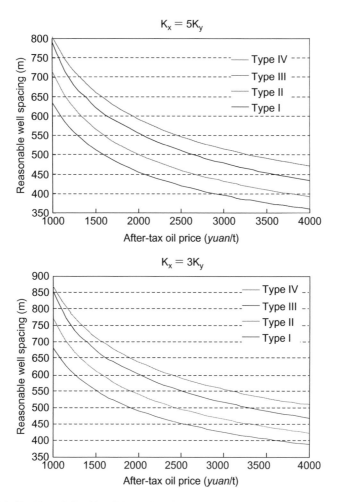

FIGURE 3-13   The relationship of the crude price to the well spacing in the diamond inverted nine-spot pattern.

for a systematic evaluation of the well patterns in the development of ultralow-permeability reservoirs, which lays a scientific basis for the deployment of well patterns.

3.1.2.4 An Example of Calculating the Reasonable Well-Spacing/ Row-Spacing Ratio (d_x/d_y)

The reasonable d_x/d_y ratio is determined in a way similar to the reasonable well density in that it can be changed either by fixing the well spacing while changing the row spacing or vice versa. Then on the basis of the ratio changes, different well patterns will be designed and their technical indexes calculated. A comparative analysis (Table 3-10) reveals the following results.

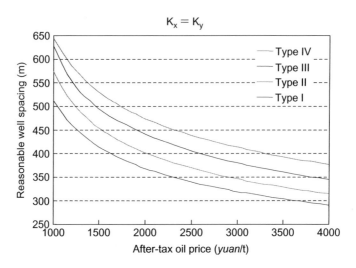

FIGURE 3-14 The relationship of the crude price to the well spacing in the diamond inverted nine-spot pattern.

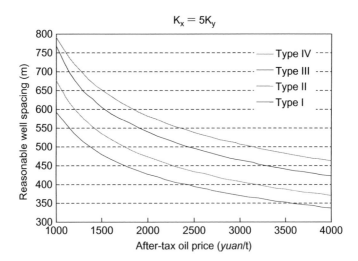

FIGURE 3-15 The relationship of the crude price to the well spacing in the interlaced-row well pattern.

In a given production period, a higher d_x/d_y ratio may bring a higher recovery, but also a higher water-cut.

Different d_x/d_y ratios may have little influence on the final recovery but may have a great influence on the duration of oilfield development, i.e., a higher dx/dy ratio may shorten the production time. Generally speaking, a higher dx/dy ratio may produce a better development result. However, this does not mean that the higher the ratio the better the result, but that there is an optimal value of the ratio, i.e., a reasonable dx/dy ratio, which

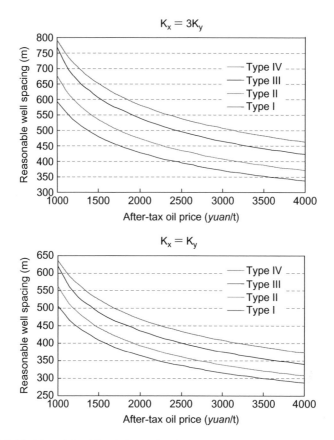

FIGURE 3-16 The relationship of the crude price to the well spacing in the interlaced-row well pattern.

can produce the maximum financial net present value (FNPV). Therefore, we need to perform an economic evaluation for the benefits of different dx/dy ratios.

Fig. 3-17 presents the relation between the accumulative FNPV and the dx/dy ratios at different stages of production. The calculation through logarithm polynomial regression according to the trend of the data points in the graph shows that the optimal dx/dy ratio is 2.7.

3.1.2.5 Adjustment of Infill Wells

Statistics of the 28 infill wells on the lateral sides of fractures in the Wangyao area of Ansai show that the infill wells bring a high initial and current output with little decline when the row spacing ranges between 80 m and 150 m, but their effect is not as good when the row spacing exceeds 150 m, and a spacing below 80 m may increase the water-cut quickly, thus producing an even poorer effect (Table 3-11).

TABLE 3-10 A Contrast of Technical Indexes of the Diamond Well Pattern with Different dx/dy Ratios in the Baima Area

dx/dy ratio		5.0	3.3	2.9	2.5
fw = 90%	η (%)	23.66	23.61	23.14	22.34
	Time (year)	20.2	30.2	33.2	39.2
fw = 95%	η (%)	27.50	27.45	26.94	27.33
	Time (year)	36.2	53.2	61.2	72.2
5 years	η (%)	13.65	10.68	9.40	8.36
	fw (%)	53.11	31.24	21.49	15.83
10 years	η (%)	18.91	15.92	14.73	13.69
	fw (%)	77.57	66.81	60.28	52.24
20 years	η (%)	23.80	20.95	19.73	18.88
	fw (%)	90.26	84.53	82.17	78.90

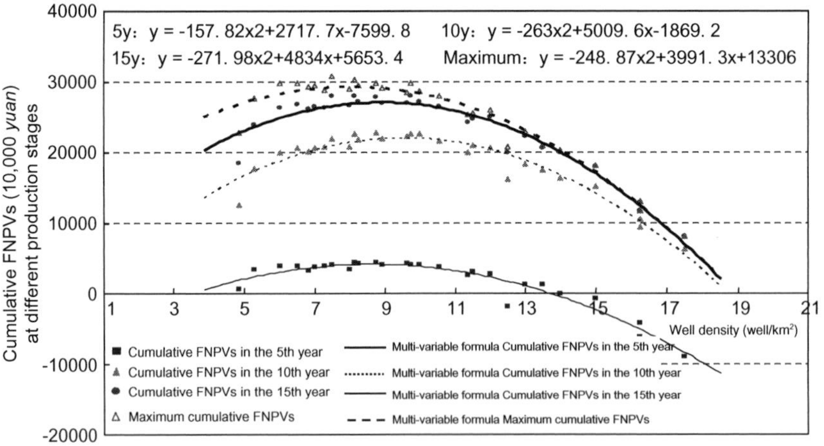

FIGURE 3-17 A contrast between the cumulative FNPVs at different production stages and the well densities of the diamond patterns with different dx/dy ratios in the Baima area.

Combining all the methods above, we decided that the reasonable row spacing should be kept between 100 m and 180 m, and the well spacing between 450 m and 550 m for developing ultralow-permeability reservoirs in Changqing.

TABLE 3-11 Production Data of Infill Wells with Different Row-Spacing Patterns in the Wangyao Area of Ansai

Row spacing (m)	Number of infill wells	Initial stage			Half a year later		
		Oil output (t/d)	Water-cut (%)	Working fluid level (m)	Oil output (t/d)	Water-cut (%)	Working fluid level (m)
Below 80	1	3.76	32.6	477	1.50	53.0	1180
80~120	14	5.07	11.6	691	3.14	16.4	959
120~150	11	4.67	12.5	657	2.90	22.8	1013
Over 150	2	2.33	6.3	329	1.45	11.9	506

3.2 TIMING OF ADVANCED WATER INJECTION

3.2.1 Pressure Distribution between the Water Injector and the Oil Producer in Ultralow-Permeability Reservoirs

3.2.1.1 Pressure Distribution between the Injector and the Producer with Stable Seepage Flow

Pressure Distribution in Areal Radial Flows

Suppose the flow of fluids in the porous media follows the rule of areal radial seepage. Then the pressure at any point in the stable areal radial seepage can be expressed as

$$p = p_e - \frac{(p_e - p_w) - G(r_e - r_w)}{\ln\left(\dfrac{r_e}{r_w} + S\right)} \ln \frac{r_e}{r} - G(r_e - r)$$

$$= p_w + \frac{p_e - p_w - G(r_e - r_w)}{\ln\left(\dfrac{r_e}{r_w} + S\right)} \ln \left(\frac{r}{r_w}\right) + G(r - r_w)$$

(3.12)

and the equation for the output of the stable areal radial seepage is

$$Q = 86.4 \frac{2\pi K K_{ro} h[(p_e - p_w) - G(r_e - r_w)]}{\mu_o B_o \left(\ln \dfrac{r_e}{r_w} + S\right)}$$

(3.13)

where

Q—per-well output, m^3/d

r_e and r_w—radius of the drainage boundary and that of the well bore respectively, m

p_e and p_w—pressure on the drainage boundary and the bottom-hole flowing pressure, respectively, MPa

G—TPG, MPa/m

μ_o—viscosity of oil in place, $mPa \cdot s$

B_o—oil volume coefficient on the ground and underground, f

Pressure Distribution between the Injector and the Producer in the Case of Fixed Production

Fig. 3-18 is a diagram of one injector, serving as the energy source, and one producer, serving as the fluid convergence. In the seepage field formed by the source and the convergence, the pressure distribution at any point in the formation with a TPG considered is different from that without a TPG considered. Suppose the potential at any point M in the formation is

$$\Phi_M = \Phi_{M1} + \Phi_{M2} \tag{3.14}$$

$$\Phi_{M1} = -\frac{Q}{2\pi h} \ln r_1 - \Phi_{s1} + C_1 = -\frac{Q}{2\pi h} \ln r_1 - \frac{K}{\mu} Gr_1 + C_1 \tag{3.15}$$

$$\Phi_{M2} = \frac{Q}{2\pi h} \ln r_2 + \Phi_{s2} + C_2 = -\frac{Q}{2\pi h} \ln r_2 - \frac{K}{\mu} Gr_2 + C_2 \tag{3.16}$$

$$\Phi_M = \Phi_{M1} + \Phi_{M2} = \frac{Q}{2\pi h} \ln \frac{r_2}{r_1} + \frac{K}{\mu} G(r_2 - r_1) + C \tag{3.17}$$

Φ_{s1} and Φ_{s2} in the formulae are, respectively, the TPG potentials of the injector W_{inj} and of the producer W_p at point M.

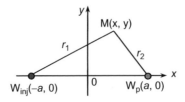

FIGURE 3-18   Diagram of one injector, the source and one producer, the convergence.

Suppose $r_1 = r_w$, $r_2 = 2a + r_w \approx 2a$. And $\Phi_M = \left(\dfrac{K}{\mu}\right) p_{inj}$. Then

$$\frac{K}{\mu} p_{inj} = \frac{Q}{2\pi h} \ln \frac{2a}{r_w} + \frac{K}{\mu} G \cdot 2a + C \tag{3.18}$$

Suppose $r_1 = 2a + r_w \approx 2a$, $r_2 = r_w$ and $\Phi_M = \left(\frac{K}{\mu}\right) p_w$. Then

$$\frac{K}{\mu} p_w = \frac{Q}{2\pi h} \ln \frac{r_w}{2a} - \frac{K}{\mu} G \cdot 2a + C \tag{3.19}$$

$$\frac{K}{\mu}(p_{inj} - p_w) = \frac{Q}{2\pi h} \ln \frac{2a}{r_w} + \frac{K}{\mu} G \cdot 4a \tag{3.20}$$

$$Q = \frac{2\pi h K (p_{inj} - p_w - G \cdot 4a)}{\mu \ln \dfrac{2a}{r_w}} \tag{3.21}$$

The pressure at any point M in the formation can be expressed as

$$\begin{aligned}
p &= p_w + \frac{Q}{2\pi kh} \ln \frac{r_2 \cdot 2a}{r_1 r_w} + G \cdot 2r_2 \\
&= p_w + \frac{(p_{inj} - p_w - G \cdot 4a)}{\mu \ln \dfrac{2a}{r_w}} \ln \frac{r_2 \cdot 2a}{r_1 r_w} + G \cdot 2r_2
\end{aligned} \tag{3.22}$$

Figure 3-19 shows that due to the existence of a TPG in an ultralow-permeability reservoir, the pressure around the injector tends to be high while that around the producer tends to be low. This is because the high flow resistance between the former and the latter prevents a timely energy supplementation. It is for this reason that advanced injection can improve such a situation to a certain extent.

Areal Pressure Distribution in Different Well Patterns

An electric simulation experiment is a physical simulation experiment based on the similarity between water and electricity, which can simulate various complex fluid flows so as to vividly get the flow dynamics of the fluids, the oil-well output, the pressure distribution, and waterline advancement. A large number of figures have been obtained about the seepage fields in the cases of production through straight-hole fracturing and through horizontal well patterns. The areal seepage fields in the two cases are consistent with each other because our experiments just simulate the conditions of infinite fracture

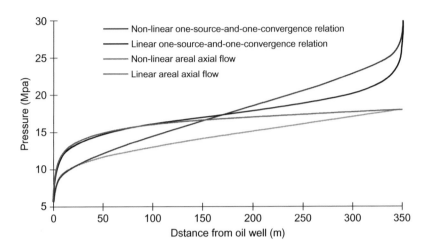

FIGURE 3-19   Pressure distribution between different drainage boundaries and production wells.

conductivity in both straight-hole fracturing and horizontal well patterns. Therefore, what is presented here is just the results from the analysis of the characteristics of pressure distribution between wells through experiments with the horizontal patterns. The well spacing and the row spacing in the well patterns to be discussed here are 720 m and 240 m, respectively.

Figure 3-20 presents the isopotential lines when both the producer and the injector in the diamond, rectangular, and interlaced-row patterns are horizontal wells. The thick solid lines in the figure represent the horizontal sections.

3.2.1.2 Pressure Distribution between the Injector and the Producer with Unstable Seepage Flow

Theoretical Studies

Suppose that there is one injector and one producer in an infinite fractured ultralow-permeability reservoir with a uniform thickness (Fig. 3-21). We can establish a rectangular coordinate system with its x-axis and y-axis representing the main trends of anisotropic permeabilities and the midpoint of the line connecting the two wells serving as the origin of the coordinates. Suppose the angle between the connecting line and the x-axis is θ, the coordinates of the injector and the producer are (x_0, y_0) and (x_1, y_1), respectively, and the distance between the origin and the two wells is a. We also suppose in our studies that the changes in fluid viscosity, porosity and permeability in the seepage flow process are not taken into account; that the injected fluid and the produced fluid have the same mobility ratio; that the injector and the producer operate at the same time; that the active area keeps expanding; and

TABLE 3-12 A Comparison of Different Advanced-Injection Programs for the Triassic Reservoirs

Program	Production stage	Per-well output (t/d)	Water-cut (%)	Injection rate (m³/d)	Formation pressure (MPa)	Oil producing rate (%)	Recovery (%)
Program 1: Injecting water 12 months after commissioning	The 1st year	2.7	7.5	0	6.3	1.20	1.81
	The 10th year	1.6	71.5	234	16.2	0.67	16.53
Program 2: Injecting while in production	The 1st year	6.4	5.3	310	9.9	2.78	2.60
	The 10th year	1.6	83.0	305	15.6	0.65	19.23
Program 3: Advanced injection at 0.48% PV	The 1st year	6.9	5.1	298	10.6	3.01	2.91
	The 10th year	1.7	82.4	279	16.1	0.72	19.82
Program 4: Advanced injection at 4.1% PV	The 1st year	7.3	5.0	280	11.4	3.18	2.97
	The 10th year	1.6	82.5	279	16.1	0.71	19.86
Program 5: Advanced injection at 5.9% PV	The 1st year	7.3	4.9	263	12.2	3.19	2.98
	The 10th year	1.6	82.5	279	16.1	0.71	19.88
Program 6: Advanced injection at 7.0% PV	The 1st year	7.3	4.8	239	12.8	3.19	2.98
	The 10th year	1.6	82.4	279	16.1	0.71	19.89

TABLE 3-13 Variations of the Minimum Crustal Stress in Production Wells Over Time

Well number	Minimum crustal stress, MPa					
	Original	Sept. 2004	Mar. 2005	Sept. 2005	Feb. 2006	Aug. 2006
Zhuang-1	32.4	32.4	33.4	35.4	35.6	35.6
Zhuang-2	35.8	35.8	35.8	35.8	36.1	37.4

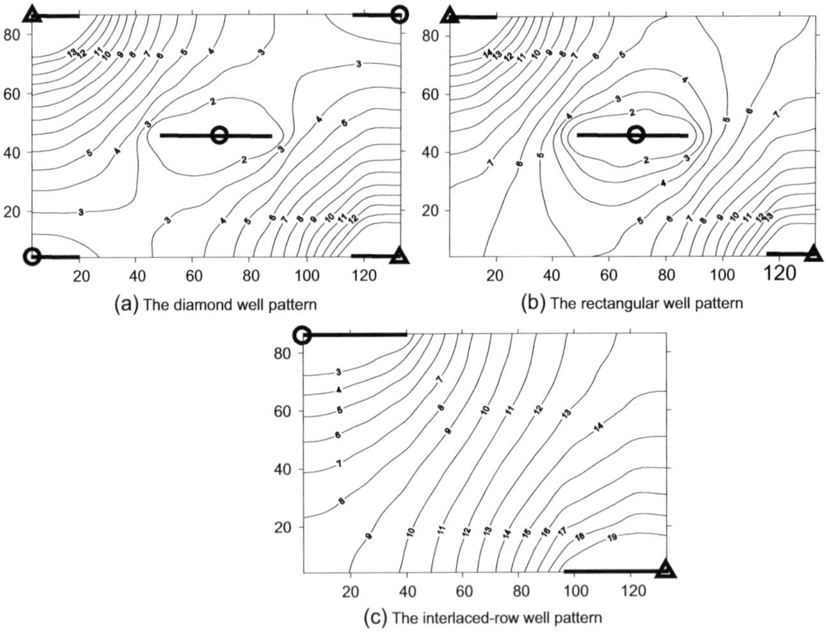

(a) The diamond well pattern

(b) The rectangular well pattern

(c) The interlaced-row well pattern

FIGURE 3.20 The seepage flow field distribution in different well patterns when the producer and the injector are both horizontal wells.

that the moving boundary of the injector encounters that of the producer at time t_0. Well interference will occur when $t > t_0$.

When $t > t_0$, the pressure distribution in a fractured anisotropic ultralow-permeability reservoir at t, according to the superposition principle, can be expressed as

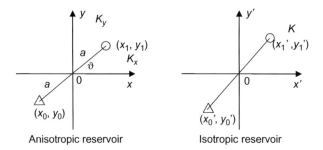

FIGURE 3-21   Well interference between one injector and one producer in a fractured low-permeability reservoir.

$$p = p_e + \frac{Q_p\mu}{4\pi\sqrt[3]{K_xK_yK_z}h\beta_3}\left[1 - \left(\frac{(x\beta_1 - a\beta_1\cos\theta)^2 + (y\beta_2 - a\beta_2\sin\theta)^2}{R_p^2(t)}\right)\right]$$

$$+ \frac{Q_p\mu}{2\pi\sqrt[3]{K_xK_yK_z}h\beta_3}\cdot\ln\frac{\sqrt{(x\beta_1 - a\beta_1\cos\theta)^2 + (y\beta_2 - a\beta_2\sin\theta)^2}}{R_p(t)}$$

$$- G_p\left[R_p(t) - \sqrt{(x\beta_1 - a\beta_1\cos\theta)^2 + (y\beta_2 - a\beta_2\sin\theta)^2}\right]$$

$$+ \frac{Q_i\mu}{4\pi\sqrt[3]{K_xK_yK_z}h\beta_3}\left[1 - \left(\frac{(x\beta_1 + a\beta_1\cos\theta)^2 + (y\beta_2 + a\beta_2\sin\theta)^2}{R_i^2(t)}\right)\right]$$

$$+ \frac{Q_i\mu}{2\pi\sqrt[3]{K_xK_yK_z}h\beta_3}\cdot\ln\frac{\sqrt{(x\beta_1 + a\beta_1\cos\theta)^2 + (y\beta_2 + a\beta_2\sin\theta)^2}}{R_i(t)} \qquad (3.23)$$

$$- G_i\left[R_i(t) - \sqrt{(x\beta_1 + a\beta_1\cos\theta)^2 + (y\beta_2 + a\beta_2\sin\theta)^2}\right]$$

$$R_i^2(t)\left[\frac{8\pi\sqrt[3]{K_xK_yK_z}h\beta_3 G_i}{Q_i\mu}\cdot R_i(t) + 3\right] = 24\eta t \qquad (3.24)$$

$$R_p^2(t)\left[\frac{8\pi\sqrt[3]{K_xK_yK_z}h\beta_3 G_p}{Q_p\mu}\cdot R_p(t) + 3\right] = 24\eta t \qquad (3.25)$$

$$R_i(t) = \sqrt{[x(t)\beta_1 + a\beta_1\cos\theta]^2 + [y(t)\beta_2 + a\beta_2\sin\theta]^2} \qquad (3.26)$$

$$R_p(t) = \sqrt{[x(t)\beta_1 - a\beta_1\cos\theta]^2 + [y(t)\beta_2 - a\beta_2\sin\theta]^2} \qquad (3.27)$$

where

K_x, K_y, K_z—principal values of reservoir permeability, $\times 10^{-3} \mu m^2$

K—average permeability in fractured reservoirs, $10^{-3} \mu m^2$

ρ—fluid density, kg/m^3

φ—reservoir porosity, frac

v —velocity vector, m/s

p—pressure, MPa

t —time, d

G—TPG, MPa/m

η—pressure-transmitting coefficient in the strata, m^2/s

μ—fluid viscosity, mP·s

C_f—comprehensive compressibility, 1/MPa

p_e—reservoir pressure, MPa

Q—oil output, m^3/d

$R(t)$—moving boundary radius, m

According to the percolation theory, the main flow line, with the highest flow velocity among all the flow lines, connects the injector with the producer under real reservoir conditions. The moving boundaries of the injector and the producer on the main flow line will meet before others. Afterwards, the two wells begin to interfere with each other.

The Interference of the Injector and the Producer with Each Other in a Fractured Ultralow-Permeability Reservoir

First, the differential permeability (i.e., the ratio of the maximum permeability to the minimum) in the Zhuang-9 well area, which is 5.29, is calculated using the equivalent continuum model of the fractured reservoir that has already been established. Second, the well interference theory is used to study how the injection-to-production ratio, injector-producer spacing, and relative relationship between the injector-producer direction and the fracture trend influence the pressure distribution. An insufficient reservoir pressure at the bottom of a producer in operation will lead to serious deaeration and increase the flow resistance around the borehole. So 7 MPa is set as the minimum flowing pressure, below which production will be stopped due to difficulty of feed flow, according to the field data.

Figures 3-22(a) and (b), respectively, show the pressure distribution curves of different injection-to-production ratios in a homogeneous reservoir after 200 days and 26 days of production (thereafter the bottom-hole flowing pressure is lower than 7 MPa) with TPGs of 0.05 MPa/m and 0.20 MPa/m, respectively. Figure 3-23 presents curves demonstrating how the TPG and injection-to-production ratio affect the bottom-hole flowing pressure (BHFP). From Figs. 3-22(a) and 3-23 it can be seen that, with a lower TPG, the pressure wave can transmit effectively between the injector and the producer, which enables the producer to make a quick response.

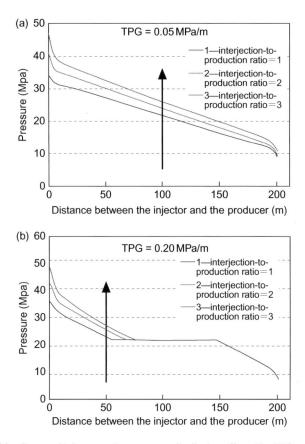

FIGURE 3-22   Curves of injector-producer pressure distribution affected by TPGs and injection-to-production ratios in a homogeneous reservoir.

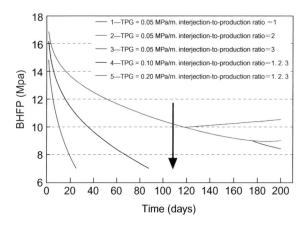

FIGURE 3-23   Curves of BHFP distribution affected by TPGs and injection-to-production ratios in a homogeneous reservoir.

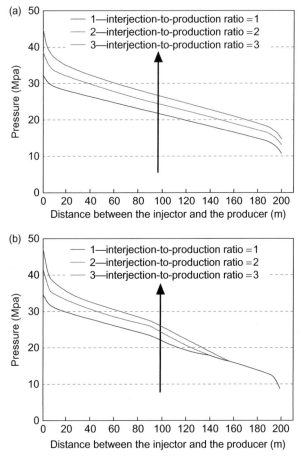

FIGURE 3-24 Pressure distribution curves along the injector-producer line in fractured reservoir (well spacing: 200 m; production period: 200 days). (a) The injector-producer line parallel to the fracture trend. (b) The injector-producer line perpendicular to the fracture trend.

Therefore, a proper rise of the injection-to-production ratio can help the reservoir pressure trigger a quicker response in the producer. According to Figs. 3-22(b) and 3-23, with a higher TPG, the feed flow becomes insufficient after 26 days of production. As the injector-producer pressure curve is shaped like a staircase in the middle, which means these sections are not covered by the pressure, raising the injection-to-production ratio can expand the pressure swept region near the end of the injector, but cannot trigger any response in the producer.

Figures 3-24 to 3-27 present curves of the pressure distribution on the injector-producer connecting line and in the bottom hole affected by the

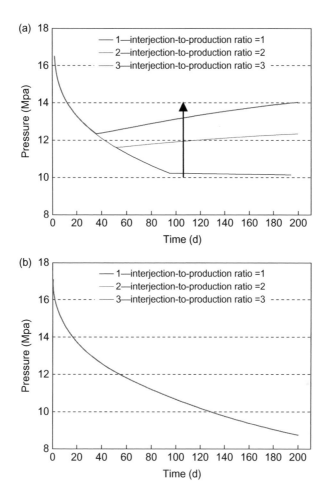

FIGURE 3-25 Curves of BHFP variations in fractured reservoirs (well spacing: 200 m). (a) The injector-producer line parallel to the fracture trend. (b) The injector-producer line perpendicular to the fracture trend.

injection-to-production ratio, the line trend, and the injector-producer spacing in the fractured reservoir where $K_x/K_y = 5.29$.

As shown in these figures, when the injector-producer connecting line is parallel to the fracture trend, a rise of the injection-to-production ratio can increase the pressure on the line and stop the BHFP drop after the bottom hole responds. Further increase of the injection-to-production ratio may enable the BHFP to rise gradually. The higher the injection-to-production ratio, the quicker the producer responses and the pressure rise. During the 200 days of production under our observation, when the injector-producer line was perpendicular to the fracture trend, the pressure near the injector

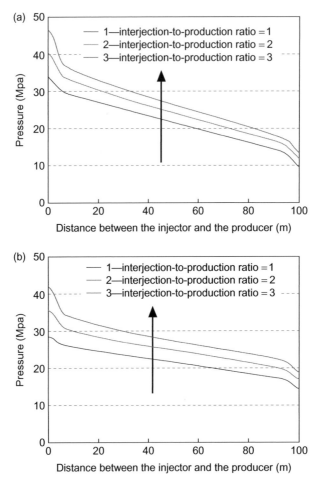

FIGURE 3-26 Pressure distribution curves along the injector-producer line in fractured reservoirs (well spacing: 200 m; production period: 200 days). (a) The injector-producer line parallel to the fracture trend. (b) The injector-producer line perpendicular to the fracture trend.

rose with the increase of the injection-to-production ratio, while the production pressure near the producer and the BHFP remained unchanged, which means the oil well had no response. Therefore, for a fractured ultralow-permeability reservoir with poor continuity, we can make full use of natural fractures by raising the injection-to-production ratio and practicing displacement along the fracture trend to improve the effect of development.

Figures 3-26 and 3-27 indicate that the injection pressure at the injector drops a little when the injector-producer distance is shortened, but the BHFP rises significantly, shortening its response time. So optimizing

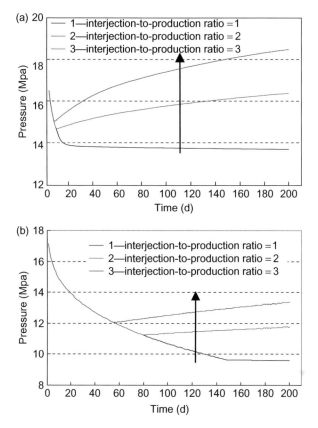

FIGURE 3-27 Curves of BHFP variations in fractured reservoirs (well spacing: 200 m). (a) The injector-producer line parallel to the fracture trend. (b) The injector-producer line perpendicular to the fracture trend.

the injector-producer spacing is the key technique for establishing an effective displacement system in both the injector and the producer.

3.2.2 Injection Timing

The worse the physical properties are in a reservoir, the higher its initial pressure will be, which requires a longer lead time of advanced injection. But raising the injection intensity appropriately can shorten the lead time (Fig. 3-28).

The injection time is related to the cumulative amount of injected water, the reasonable injection intensity, and the economic benefits. In the development of low- and ultralow-permeability reservoirs through advanced injection, the increase of injection time and cumulative water amount may cause

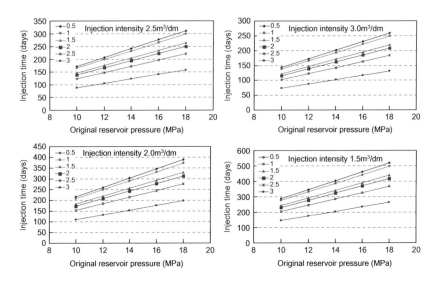

FIGURE 3-28 The relationship between the injection timing and the initial reservoir pressure and permeability.

the curve of the per-well output to go upward. But due to restrictions from various factors, such as the injection equipment, the reservoir conditions, the crude properties, the well patterns, the well densities, and the costs of advanced injection, it is impossible to extend the injection time without a limit, which means there must be a reasonable and critical timing scheme for advanced injection. From an economic perspective, the former refers to the scheme that can produce maximum economic benefits, while the latter refers to one that leads to a total profit of zero.

According to the material balance equation in an unsaturated reservoir, we can derive the equation of the cumulative oil output as

$$N_p = \frac{(N \times B_{oi} \times Ce + k)[a \times \ln(t + 1) \times p_i - p]}{B_o + B_w \times \lambda} \tag{3.28}$$

Suppose the oil output obtained through advanced injection is $(1 + \beta)$ (β generally ranging from 30% to 40%) times that of synchronous flooding. The additional output from advanced waterflood can be calculated with the following equation:

$$\beta N_p = \frac{\beta(N \times B_{oi} \times C_e + k)[a \times \ln(t + 1) \times p_i - p]}{B_o + B_w \times \lambda} \tag{3.29}$$

According to the price of crude oil and the investment in waterflooding, when the difference between the additional output and the investment M (varying with the time t of advanced injection) reaches the maximum, the corresponding time (t) turns out to be the reasonable advanced injection time; hence we obtain the following calculation equation:

$$t = \frac{\beta \times \alpha \times \eta \times p_i \times (N \times B_{oi} \times C_e + k)}{b \times (B_o + B_w \times \lambda)} \tag{3.30}$$

where

N— geologic reserves of crude oil, t

C_e—elastic compressibility in the reservoir, 1/MPa

N_p—cumulative crude oil output, t

p—current formation pressure, MPa

k—water-cut coefficient, m^3/(MPa · d)

λ—average water-oil ratio, $K = Wp$

η—crude oil price, $yuan/t$

a—pressure maintenance coefficient, obtained through simulation, $a = p'/[\ln(t+1)p_i]$, where p' refers to the formation pressure maintained through advanced flooding, MPa

b—daily investment of advanced injection, $yuan/d$

As the time t for advanced injection varies, when the value of additional crude oil produced equals the investment in advanced injection (varying with t), it turns out to be the critical advanced injection time t:

$$\beta\eta(N \cdot B_{oi} \cdot C_e + k) \cdot [a \cdot \ln(t+1) \cdot p_i - p] = b \cdot (B_o + B_w \cdot k) \cdot t \tag{3.31}$$

In the case that the price of crude oil produced from the Baima and Dongzhi areas of the Xifeng Oilfield lies between 1000 and 2200 yuan/t, then the reasonable lead time for advanced injection in the two areas is 4.6~10.1 months and 6.4~14.2 months, respectively, while the critical lead time is 9.9~31.9 months and 20.0~56.5 months, respectively.

3.3 INJECTION-PRODUCTION PARAMETERS

3.3.1 The Injection Pressure

The injector wellhead pressure is related to the BHFP, the pressure losses from oil tube friction and water nozzles, and the bottom-hole fluid level:

$$p_f = p_{wf} + p_{tl} + p_{mc} - \frac{H}{100} \tag{3.32}$$

where

p_f—the highest input pressure into the injector, MPa

p_{wf}—the BHFP in the injector, MPa

p_{tl}—pressure loss from oil tube friction, MPa

p_{mc}—pressure loss from water nozzles, MPa

H—depth of the injector, m

The input pressure is restricted mainly by the formation fracture pressure. In order to prevent formation breakdown from causing the injected water to flow along the fractures, the maximum flowing pressure into the injector is often kept below 90% of the formation fracture pressure.

3.3.2 Determination of the Injection Rate

The definition of formation compressibility is

$$C_t = -\frac{1}{V}\frac{\Delta V}{\Delta p} \tag{3.33}$$

For the formation pressure recovery, different physical properties entail different objectives. Under the conditions of the same physical properties, different formation pressures involve different differential pressures to maintain them at the same level. Therefore, the pore volume multiple of the water to be injected is under the dual control of both physical properties and formation pressures. The plate of theoretical calculation is shown in Fig. 3-29.

Numerical simulation (permeability: $2.0 \times 10^{-3}\,\mu m^2$, initial formation pressure: 12 MPa) shows that the per-well outputs, the recovery, percentage and the effects of development through an injection rate of 0.48%PV, 4.1% PV, 5.9%PV, and 7.0%PV are not very different. Therefore, we decide that an injection rate of 0.48%PV is proper for this reservoir, which is roughly consistent with the result from theoretical calculation, i.e., 0.41%PV.

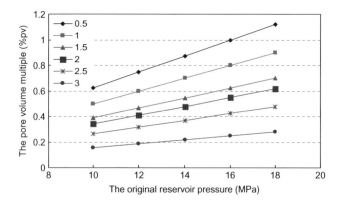

FIGURE 3-29 Relations between the injected pore volume multiple under different permeabilities and initial formation pressures.

3.3.3 Determination of the Injection Intensity

Considering the influence of TPG the injection intensity equation for the diamond inverted nine-spot pattern is

$$\frac{Q_i}{h_i} = \frac{0.5429\,K\left[(p_H - \bar{p}) + \lambda\left(0.610\sqrt{A} - r_w\right)\right]}{B_w \mu_w \left(\ln \dfrac{0.610\sqrt{A}}{r_w} - 3/4\right)} \tag{3.34}$$

and the area covered by the diamond inverted nine-spot pattern is

$$A = 4ab \tag{3.35}$$

The maximum injection intensity can be calculated by putting the previously obtained maximum injection pressure into this equation.

But from a microscopic point of view, in a water-wet reservoir, a low injection rate is good for the water behind the water-oil front to be imbibed from the high-permeability zone to the low-permeability one so as to improve the volumetric conformance efficiency.

An analysis of the effect of advanced injection in the Changqing Oilfield in recent years (Fig. 3-30) shows that the injection intensity should be kept under $3.0\ \mathrm{m^3/d \cdot m}$, with the per-well rate no more than $50\ \mathrm{m^3/d}$, as an intensity over $3.0\ \mathrm{m^3/d \cdot m}$ tends to create a fast water breakthrough and thus a quick rise of the water-cut soon after an oil producer is put into operation.

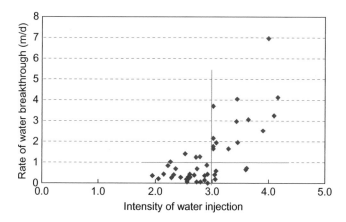

FIGURE 3-30 Relationship between the intensity of water injection and the rate of water breakthrough.

3.3.4 Reasonable Flowing Pressure in the Producer

3.3.4.1 Determining the Minimum Reasonable Flowing Pressure in the Oil Wells According to the Saturation Pressure

Thanks to the low productivity index (PI) of the oil wells in ultralow-permeability reservoirs, the flowing pressure should be kept low while the pressure drawdown is increased to maintain the wells' outputs. However, for reservoirs with relatively high saturation pressure, the degassing radius of the oil well will extend and the conditions for liquid flow in oil layers and the bore hole will become worse if the pressure drawdown is too much lower than the saturation pressure, which will affect the normal operation of the oil well.

A statistical analysis of 80 oil wells put into production in 1998 in the Chang-6 reservoir in the Wuliwan area, including classification of their data points and drawing of the scatter diagram reveal a strong regularity, with the coefficient of correlation reaching up to 0.7671. The PI reaches a maximum when the flowing pressure lies between 4 MPa and 5 MPa, which is equivalent to 60%~70% of the saturation pressure in the Chang-6 reservoir. The PI declines significantly when the flowing pressure is either too low or too high, that is to say, the flowing pressure in the producing wells at the initial stage should not be lower than 60% of the saturation pressure.

3.3.4.2 Determining the Minimum Reasonable Flowing Pressure in the Oil Wells According to the Pump Efficiency Required for Maximum Production

A too low flowing bottom hole pressure in the oil well will affect the pump efficiency. To ensure the pump efficiency, a certain pressure should be maintained at the pump inlet. The relationship between the pump efficiency and the pump intake pressure under different flowing pressures can be expressed as

$$N = \frac{1}{\left[\left(\frac{R}{10.197P_p} - a\right) + B_t\right] \times (1 - f_w) + f_w} \quad (3.36)$$

where
  N—pump efficiency
  a—solubility factor of natural gas, $m^3/(m^3 \cdot MPa)$
  P_p—pump intake pressure, MPa
  f_w—composite water-cut
  R—gas-to-oil ratio, m^3/t
  B_t—oil volume factor under the pump intake pressure

3.3.4.3 Influence of TPG and Medium Deformation on the BHFP

The curves in Figs. 3-31 and 3-32, respectively, describe how TPG and medium deformation influence the BHFP. As time passes, the BHFP during

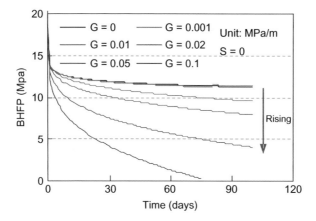

FIGURE 3-31   The effect of TPG on the BHFP.

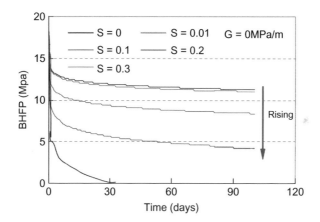

FIGURE 3-32   The effect of medium deformation on the BHFP.

fixed-output operation keeps falling, far more quickly at the initial stage than in later stages of production. With the TPG and the stress-sensitivity coefficient increasing, the BHFP will fall even lower within the same period of time, thus shortening the time of rated output. After a certain period of time, the oil well will switch from fixed-output operation to fixed-pressure operation.

3.3.4.4 The Effect of TPG and Medium Deformation on the Inflow Performance of Oil Wells

The ultralow-permeability reservoirs in China, usually under low pressure and containing little dissolved gas, are developed mainly through waterflooding.

Figure 3-33 is a comparison of inflow performance in oil wells under different seepage flow conditions. Curves 1 and 5 are variations of daily oil outputs under the condition of Darcy flow; Curves 2 and 3, respectively, take into account the influence of the deformation of media only and that of the TPG only; and Curves 4 and 6 result from the combined influence of both factors. It can be seen that under the same pressure, the daily oil output reaches the highest under the condition of Darcy flow, and the lowest under a combined influence of both TPG and medium deformation.

A look at Curves 4 and 6 together reveals that the inflow performance curve experiences a steep drop when the formation pressure drops from 20 MPa to 15 MPa. In other words, the oil output will decline when the formation pressure drops, the TPG increases, and the medium deformation is serious (Fig. 3-34).

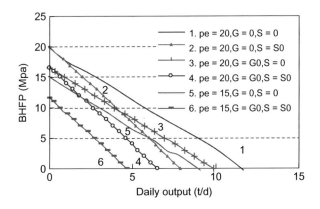

FIGURE 3-33   The influence of TPG and medium deformation on inflow performance.

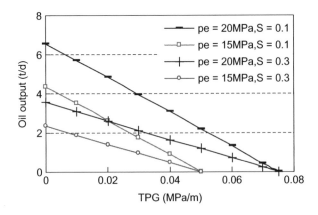

FIGURE 3-34   The influence of formation pressure, TPG, and medium deformation on inflow performance.

3.4 FRACTURING TIMING

Advanced injection, well-pattern optimization, and fracturing stimulation are the three main closely related techniques for the development of Changqing, a typical ultralow-permeability oilfield.

Advanced injection can, on the one hand, alleviate the stress damage in the reservoir, maintain a higher reservoir permeability, reduce the output decline, and increase the ultimate recovery percentage. On the other hand, it will create changes in the stress field, which dominate the extension of hydraulic fractures. For example, at the initial stage of pilot advanced flooding, fracture engineering began in the Chang-8_2 reservoir in well Zhuang-XX after four months of waterflooding. Downhole monitoring shows that the fracture height in the well is 135 m, exceeding the top and bottom barrier beds in the Chang-8_2 reservoir in the vertical direction and developed below it, which means the engineering work fails to reach its expected objectives.

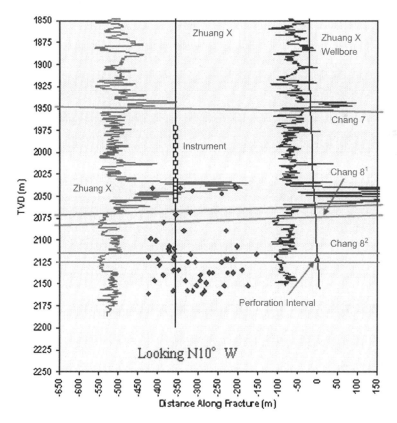

FIGURE 3-35 The monitoring result of seismic microfractures in the downhole of the Zhuang-XX well.

In the development of oilfields, especially through advanced injection, there is often a strong solid-liquid coupling effect because of the mutual influence of the stress field and the flow field on each other. Therefore, a good knowledge of the 3D variations of the crustal stress, its influence on the extension of hydraulic fractures, and the fracturing timing may help in planning appropriate measures for optimal production by exploiting or transforming the stress field to alleviate its damage.

Dynamic Distribution of the Crustal Stress in the Process of Advanced Injection

The process of advanced injection and production is actually a question of coupling between the reservoir deformation and the fluid flow. In this process, the fluid flows in the reservoir and the pressure changes will cause reservoir deformation, mainly reflected in the changes in the two principal stresses in the horizontal direction. The vertical stress will not be influenced by advanced injection and oil production because the ground, with no vertical restrictions, may get deformed freely. Variations of the crustal stress in the reservoir will change the seepage parameters in the reservoir, such as permeability, porosity, and compressibility, and then further affect the fluid flow patterns. It is clear that the two processes, i.e., the variations of the stress and the fluid flow, go on side by side and should be handled at the same time. Therefore, it is necessary to establish finite element equations for both the seepage flow field and the stress field, as follows:

$$\frac{\partial}{\partial x}\left(\rho_o \frac{kk_{ro}}{\mu_o}\frac{\partial p_o}{\partial x} - G_0\right) + \frac{\partial}{\partial y}\left(\rho_o \frac{kk_{ro}}{\mu_o}\frac{\partial p_o}{\partial y} - G_0\right) = \frac{\partial}{\partial t}(\rho_o \Phi S_O)$$

$$\frac{\partial}{\partial x}\left(\rho_w \frac{kk_{rw}}{\mu_w}\frac{\partial p_w}{\partial x}\right) + \frac{\partial}{\partial y}\left(\rho_w \frac{kk_{rw}}{\mu_w}\frac{\partial p_w}{\partial y}\right) = \frac{\partial}{\partial t}(\rho_w \Phi S_W)$$

$$S_o + s_w = 1.0$$
$$\overline{P} = P_W S_W + P_O S_O$$

where
P_o—reservoir oil pressure
P_w—reservoir water pressure
ρ_o—reservoir oil density
ρ_w—reservoir water density
S_o—reservoir oil saturation
S_w—reservoir water saturation
ϕ—reservoir porosity
k—absolute reservoir permeability

k_{ro}, k_{rw}—relative permeabilities of oil phase and water phase in the reservoir

$\overline{P}$ — average pressure of reservoir oil and water

In the process of water injection and development, the deformation of rocks and the flow of the fluids in the reservoir interact with each other. The fluids and solids (rocks) should be regarded as an overlapping continuum, for their prominent feature is that the solid zone and the fluid zone, interrelated and interactive with each other, can hardly be distinguished. This feature, therefore, requires the control equation for the fluid-solid coupling problem to target specific controlling phenomena and to reflect the fluid-solid coupling state. That is to say, the control equation describing the movement of fluids should include the factors that reflect the influence of rock deformation, while the control equation describing the movement or balance of solids should include the factors that reflect the influence of fluid movement. The equation for solid balance is

$$\begin{cases} \dfrac{\partial(\sigma_{xx} - \alpha p)}{\partial x} + \dfrac{\partial \sigma_{xy}}{\partial y} + \dfrac{\partial \sigma_{xz}}{\partial z} + f_x = 0 \\[2mm] \dfrac{\partial \sigma_{xy}}{\partial x} + \dfrac{\partial(\sigma_{yy} - \alpha p)}{\partial y} + \dfrac{\partial \sigma_{yz}}{\partial z} + f_y = 0 \\[2mm] \dfrac{\partial \sigma_{xz}}{\partial x} + \dfrac{\partial \sigma_{xy}}{\partial y} + \dfrac{\partial(\sigma_{zz} - \alpha p)}{\partial z} + f_z = 0 \end{cases}$$

In fact, medium deformation and fluid flow change side by side over time, which means that variations of the reservoir pressure may cause changes in the crustal stress. For this reason, a proper and efficient coupling method is needed. Our studies show that the time-ordered coupling can not only ensure a high analytic precision, but also speed up problem solving (Fig. 3-36).

The state of the crustal stress determines the extension and geometric shape of hydraulic fractures. For oilfields undergoing advanced injection, this is reflected in the following two respects.

On the one hand, the direction of maximal principal stress changes as water injection goes on. This may affect the efficiency of fracture engineering in that the change of the directions of hydraulic fractures may distort the fractures around the borehole while fracturing the horizontal well. For example, after advanced injection was initiated in block Zhuang-X, the minimum principal stress increased significantly over time at some points with high injection intensities. The maximal principal stress in this area was originally distributed in the main direction of 40−70 northeast (Fig. 3-37). But it had turned 5−10 by September 2005 (Fig. 3-38) and 15 by August 2006 (Fig. 3-39), clockwise relative to the original direction (Figs. 3.36 to 3.39). At the same time, the local maximum principal stress direction in some areas turned counterclockwise.

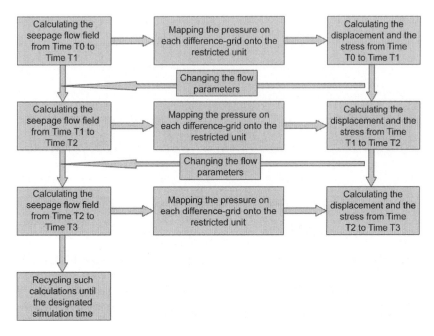

FIGURE 3-36   Procedures for the coupling method.

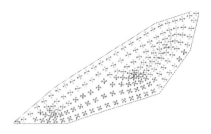

FIGURE 3-37   The direction of the original crustal stress.

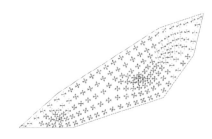

FIGURE 3-38   The direction of the crustal stress in Sept. 2005.

FIGURE 3-39   The direction of crustal stress in Aug. 2006.

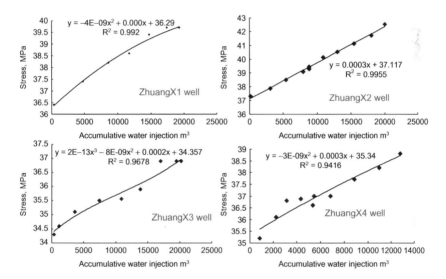

FIGURE 3-40   Variations of the minimum crustal stress in production wells over time.

On the other hand, during advanced injection in ultralow-permeability reservoirs, the horizontal stress increases with the increase of the pore pressure in the oil layer, while the barrier pressure and its horizontal stress remain approximately the same. As a result, the stress difference tends to drop with the increase of injection time and rate. The data of separate-layer crustal stress is the key to determining the fracture height. Therefore, it becomes of vital importance to study the stress difference caused by the extension of hydraulic fractures in the oil layer only on the one hand and the relationship between the height that hydraulic fractures get into the barrier and the stress difference on the other. Take the flooded Chang-8 reservoir in the Zhuang-X well block, for example. Its minimum crustal stress goes up gradually with the cumulative amount of injected water increased (Fig. 3-40).

The minimum crustal stress has a positive correlation with the cumulative injection rate, which can be expressed through straight lines, curves, or power functions.

The Best Time for Fracturing in Low-Permeability Reservoirs

Advanced injection that lasts a certain period of time may change the direction of the crustal stress, thus causing changes in the fracture direction and probably early water breakthroughs in the oil well. In addition, it may also increase the reservoir stress, thus reducing the stress difference between the barriers and the oil layers, and probably making the height of hydraulic fractures difficult to control and causing hydraulic fracturing to fail as a whole. Is it possible for artificial fracturing to start before advanced injection? It is well known that, as the flow conductivity of propping fractures created by hydraulic fracturing may decrease gradually over time, the oil-well productivity will also decline. It is clear that hydraulic fracturing before advanced injection may prolong the time when the propping fractures bear pressure. Meanwhile, the gradual increase of the crustal stress in the process of advanced water injection, which means that the hydraulic fractures have to bear an increasing force, may cause conductivity to drop more quickly. As a consequence, the oil producer does not operate while the fractures have a higher conductivity, which has already declined when it is opened. From this point of view, hydraulic fracturing should begin after water injection in order to maximize the efficiency of artificial fractures and get high oil outputs.

Therefore, the best time for fracturing should be based not only on the stress difference between the reservoir and the barrier, but also on the decrease of fracture conductivity, so as to maximize the overall benefits.

With the advanced injection progressing, the minimum horizontal ground stress increases by $2-3$ MPa, which means that the stress difference between the reservoir and the barrier drops by $2-3$ MPa correspondingly. This is very unfavorable for the control of the fracture height. Therefore, the reservoir-and-barrier stress difference is the main factor affecting the best fracturing time.

Laboratory physical and mechanical simulations are carried out to study relations between the extension of hydraulic fracture height and the stress difference. Physical simulations are carried out by imposing different stresses on man-made coupons to find out the critical stress difference that can enable the artificial fracture to enter the barrier from the oil layer.

On the basis of the physical experiments, a mechanical model of fracture propagation is established under the required conditions, i.e., the intensity factor of the stress field K_I around the fracture terminal is equal to its critical value K_{IC}. The relation between the distance for fractures to enter the barrier and the stress difference is gained by changing the stress in different layers in the physical and mechanical model.

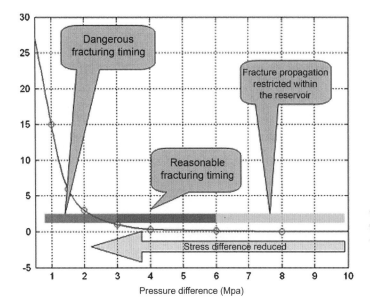

FIGURE 3-41 Relationship between the distance for the fracture to enter the barrier and the stress difference.

The stress difference can be divided into three stages based on the results from calculations and simulations of for fractures to enter the barrier.

The distance for fractures to enter the barrier tends to increase as the stress difference falls. Fracturing design should take into account both the stress difference changes and the thickness of oil layers and barriers in the development of oilfields through advanced injection and try to do the fracture job within the range of stress difference that corresponds to the reasonable fracturing time (Fig. 3-41).

Take, for instance, the diamond inverted nine-spot well pattern in the Zhuang-X area. If composite high-energy perforated injection is carried out at a rate of 30 m^3/d, then the best time for advanced injection is when the formation pressure rises up to 110% of its original value; the injection work under this pressure should last over three months.

Chapter 4

Technical Support for Advanced Water Injection

Advanced water injection has become an efficient and distinctive technology for developing ultralow-permeability reservoirs. Compared with delayed and synchronous injection, advanced injection has a higher demand for supporting projects, not only including adequate studies of early evaluation and an overall framework planning of water injection before production, but also involving a whole process management of injection modes and methods. In our practice in the Changqing Oilfield, the systematic advanced injection program acquires the features of "quick evaluation to set up its basis, overall planning to define its scale, innovative facilities to facilitate its practice, and nodal control to promote its efficiency." In the zones of proven reserves, "Three In-advances" and "Three Priorities" have been summarized and applied. "Three In-advances" means predicting the development scale in advance, building a water injection system in advance, and

Advanced Water Injection for Low Permeability Reservoirs.
© 2013 Petroleum Industry Press. Published by Elsevier Inc. All rights reserved.

constructing a water supply system in advance. "Three Priorities" means priorities should be given to water injector drilling, water injection pipeline construction, and investment in water injection before commissioning. In the zones of unproven reserves, the mobile water injection skid integrated with a smart flow-regulating valve complex is applied. Thanks to the employment of all these measures and techniques, advanced water injection works well on a large scale.

4.1 TECHNIQUES FOR QUICK EVALUATION OF LARGE ULTRALOW-PERMEABILITY RESERVOIRS

The Ordos. Basin, characteristic of large ultralow-permeability lithologic reservoirs, has favorable formation conditions for practicing advanced injection. In order to find relatively high-permeability zones in these reservoirs and avoid stripper wells during capacity building, the techniques for quick evaluation of reservoirs and well productivity serve as not only the basis for but also the key to the practice of advanced injection.

Quick evaluation in large reservoirs includes evaluation of reservoirs and well productivity to identify the oil/gas-rich zones by analyzing the factors that influence productivity, constructing productivity-sensitive parameters, selecting evaluation parameters that combine various methods and parameters, and establishing criteria for classified evaluation. The purpose of a quick reservoir evaluation is to facilitate advanced injection and construct the ground system in advance. At the same time, such evaluation can minimize the risks of drilling low efficiency wells and dry wells, thus providing a guide to the overall strategy and improving the efficiency in the development of large low permeability lithologic reservoirs in the Ordos Basin.

4.1.1 Techniques for Quick Evaluation

4.1.1.1 Classified Evaluation of Reservoirs

Evaluation of ultralow-permeability reservoirs should focus on their effective capacities to catch oil and their permeabilities. In other words, porosity and permeability are two macroscopic parameters needed for describing the physical properties of a reservoir. On the other hand, microscopic parameters are obtained based on the data received from mercury injection and image analysis. Nuclear magnetic resonance (NMR) provides us with the parameter of movable fluid saturation, which can help get the amounts of movable fluids in place and thus make reservoir evaluation more precise.

With reference to the data about petrofacies, physical properties, pore structures and movable fluid saturations, the domestic and international criteria for reservoir classifications, and the real conditions in place, the ultralow-permeability reservoirs in the Changqing Oilfield can be classified into the types shown in Table 4-1.

TABLE 4-1 Classification of Low-Permeability Sandstone Reservoirs of the Yanchang Group

Type	Lithology	Pore structure			Physical properties		Percentage of movable fluids
		Pd	γ_{50}	σ	Porosity %	Permeability × $10^{-3} \mu m^2$	
I	Medium-to-fine sandstones	<0.35	>0.5	>2	>15	>3	>50
II	Fine sandstones	0.35~0.65	0.15~0.5	2~2.5	10~15	1~3	30~50
III	Fine-silt sandstones	0.65~1.0	0.1~0.15	1.5~2.0	8~12	0.2~1.0	20~30
IV	Siltstone	>1.0	<0.1	<1.5	<8	<0.2	<20

A unified criterion for evaluating ultralow-permeability reservoirs allows for comparison between them across the entire basin and lays a foundation for evaluating and selecting the capacity-building targets and rapidly predicting their productivity. On this basis, graphics for tight lithologic reservoir identification have been made for different development units. With reference to differences in sedimentation, diagenism, reservoir characteristics, and the "four-property" relationship, graphics for log interpretation and minimum electricity for economical development have been made for finely-divided interpreting units, covering 35 bed sets in different blocks. The online interpretation of these graphic data improves the accuracy of oil-layer interpretation and the efficiency of on-the-spot identification.

4.1.1.2 Quick Evaluation of Well Productivity

The many methods used to evaluate and predict per-well productivity can be divided into two categories. One is static-parameters evaluation based on log information, and the other is dynamic-parameters evaluation based on well-testing data. The establishment of graphics for fine-log interpretation and criteria for minimum electricity in economical development provides a theoretical basis for a preliminary prediction of per-well productivity through static-parameters evaluation based on log information, which is the focus of our discussions here. On the basis of various reservoir parameters acquired from log processing, we can analyze the relations between these parameters and reservoir productivity and find out those closest to reservoir productivity (including both qualitative and quantitative parameters) to predict the productivity through the quickest and most vivid and accessible methods, such as morphological differences of logging curves, sandbody structures, and weighted storativity.

Methods for Quick Evaluation

Induction from Logging Curve Shapes Studies have shown that the logging curves of sandbody structures and their shapes and matching can also vividly disclose the reservoir productivity, thus establishing an integrated method of logging curve induction, i.e., analysis of logging curve shapes, differentiation of three porosity curves, and comprehensive evaluation of sandbody structures.

Analysis of Logging Curve Shapes Take the Chang-6 reservoir of Huaqing as an example. There is a close relationship between the testing productivity and the logging curve shapes in different reservoirs, as shown in Table 4-2.

In the Chang-6 reservoir in well Yuan-432, the gamma-ray curve appears like a box, the resistivity curve is like a plump box that presents high values, while the curves of interval transit time and the density reveal good physical properties. These features show that this is a typical massive reservoir. Well testing obtains a high productivity of 53.6/d (Fig. 4-1).

TABLE 4-2 Logging Curve Features of Different Reservoirs in the Chang-6 Reservoir of Huaqing

Curve shapes (RILD GR R4 AC)	Shape description	Sedimentary features	Electric parameters	Sample log	Testing production (t/d)
	The resistivity curve is like a plump box, the gamma-ray curve like a standard box, the 4m resistivity goes upward steeply on the top of the oil layer and the acoustic-wave curve extends gently and indicates high values.	Body-of-river sedimentation, with high energy of hydraulic dynamics. The sediments are large-sized and evenly distributed. The material supply is adequate while the sedimentation is quick.	$Rt>35\Omega.m$, $AC>225\mu s/m$, $K>0.35mD$, $\Phi>11.0\%$ $H>17m$	The grade above oil patches, evenly distributed.	$Q>20$
	The resistivity curve is shaped in bell, hopper, box or a combination of these, and the 4m resistivity goes upward at the bottom of the oil layer.	Medium-to-low-energy sedimentary environment, with large sediments overlying fine ones, presenting a sedimentary sequence of lateral migration of the river channel and a structure of normal grain-size sequence.	$28<Rt<35\Omega.m$, $218<AC<225\mu s/m$, $0.2<K<0.35mD$, $10.5\%<\Phi<11.0\%$	The grade above oil stains and patches, evenly or unevenly distributed.	$10<Q<20$
	Both the resistivity curve and the curve of interval transit time zigzag and change in a wide range, revealing a strong heterogeneity of the reservoir.	Medium-to-low-energy sedimentary environment, with relatively fine sediments, which means an inadequate material supply.	$24<Rt<28\Omega.m$, $212<AC<218\mu s/m$, $0.08<K<0.2mD$, $8\%<\Phi<10.5\%$	The grade above oil stains, unevenly distributed.	$4<Q<10$

TABLE 4-3 Statistics of Improved Interpretation Results from Exploration and Appraisal Wells in 2009

Serial No.	Well No.	Layer	Perforated section (m)	Primary interpretation	Fine interpretation	Oil testing result	
						Oil output (t/d)	Water output (m³/d)
1	Chi-89	Chang-4+5$_2$	2279.0~2282.0	Poor oil layer & oil-water layer.	Oil layer.	20.32	0.00
2	Huang-163	Chang-6$_1$	2529.5~2531.5 2539.5~2541.5	Oil-water layer.	Oil layer.	20.15	0.00
3	Hu-191	Chang-7$_2$	2172.0~2174.0 2182.0~2186.0	Oil-water layer.	Oil layer.	14.37	0.00
4	Hu-193	Chang-7$_2$	2175.0~2179.0	Poor oil layer.	Oil layer.	12.84	0.00
5	Yan-205	Yan-8	2046.0~2047.0	Oil-water layer.	Oil layer.	10.63	0.00
6	Chi-91	Chang-4+5$_2$	2521.0~2524.0	Poor oil layer.	Oil-water layer.	10.88	2.70
7	Li-168	Chang-8$_1$	2486.5~2489.0	Poor oil layer	Oil layer.	7.91	0.00
8	Zhen-297	Yan-10	1939.5~1941.0	Oil-water layer.	Oil layer.	6.80	0.00
9	Chi-222	Chang-8$_1$	2592.0~2594.0 2608.0~2610.0	Poor oil layer.	Oil layer.	6.55	0.00
10	Zhen-298	Chang-3$_1$	2012.0~2015.0	Oil-water layer.	Oil layer.	6.21	0.00
11	Zhen-302	Yan-10	1889.0~1891.0	Oil-water layer.	Oil layer.	6.04	0.00
12	Chi-228	Chang-8$_1$	2543.0~2547.0	Poor oil layer.	Oil layer.	5.78	0.00
13	Huang-138	Chang-8$_1$	2812.0~2815 .0 2824.0~2827.0	Poor oil layer.	Oil layer.	5.70	0.00
14	Zhen-500	Chang-8$_1$	2283.0~2285.0 2290.0~2291.0	Poor oil layer.	Oil layer.	5.10	0.00
15	Huang-168	Chang-8$_1$	2599.0~2602.0	Poor oil layer.	Oil layer.	4.85	0.00
16	Bai-148	Chang-6$_3$	2448.0~2452.0	Poor oil layer.	Oil layer.	4.68	0.00
17	Gao-201	Chang-10$_1$	2178.0~2180.0	Oil-bearing water layer.	Oil-water layer.	4.59	2.90
18	Hou-104	Chang-10$_1$	1943.0~1946.0	Oil-water layer.	Oil layer.	4.17	0.00
19	Gao-69	Chang-10$_1$	1831.0~1833.0	Water layer.	Oil-water layer.	4.50	3.36
20	Xing-212	Chang-10$_1$	2871.0~2874.0	Dry layer.	Oil-water layer.	2.55	4.60

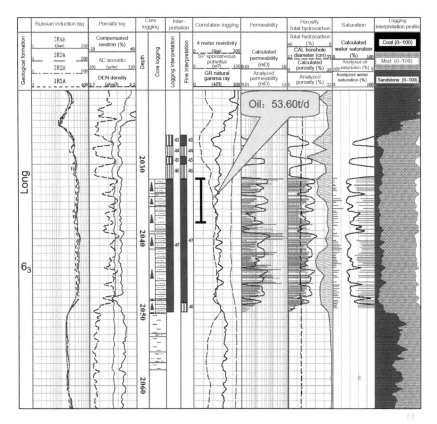

FIGURE 4-1   Log interpretation results from the Chang-6 reservoir in well Yuan-432.

In the Chang-6 reservoir in well Yuan-433, both the gamma-ray and resistivity curves are bell-shaped, with the physical properties improving as the resistivity declines, which indicates an unsaturated Type-II reservoir. Well testing obtains a commercial productivity of 17.6 t/d (Fig. 4-2).

Differentiation of Three Porosity Curves Differences between the three porosity curves, to some extent, reflect the shale contents and physical properties of reservoirs. An analysis of the relations between these differences and the testing productivity shows that in the standard calibration, the coincidence of the three porosity curves, with the reservoir densities differing largely from those of the confining rocks, indicates a Type-I reservoir and generally a high testing productivity (Fig. 4-3a). If the compensated neutron log curve coincides with the interval-transit-time log curve but not with the densilog curve, and the reservoir densities are little different from those of the confining rocks, it may mean a Type-II reservoir, which is relatively tight (Fig. 4-3b).

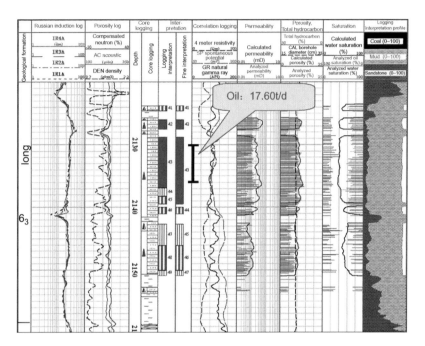

FIGURE 4-2   Log interpretation results from the Chang-6 reservoir in well Yuan-433

If the three porosity curves do not overlap, and there are small differences between the densities of the reservoir and those of the surrounding rocks, it may mean a Type-III reservoir, which is rather tight, with a testing productivity of $4 \sim 10$ t/d (Fig. 4-3c).

Comprehensive Evaluation of Sandbody Structures Studies in Jiyuan show that the sandbody structures and reservoir controlling factors in the Chang-4 + 5 oil layers are rather complex. According to sandbody types and oil-bearing features, reservoirs can be divided into four categories:

- Massive reservoirs in superposed distributary channels/mouth bars
- Composite reservoirs in dispersed distributary channels/mouth bars
- Bottom-water reservoirs in superposed distributary channels/mouth bars
- Oil-water layers in dispersed distributary channels/mouth bars

Massive reservoirs. In these layers the distributary channels are overlaid by mouth bars, with the layer thickness usually more than 10 m. Such reservoirs have good barriers both on the top and at the bottom, containing few interlayers, and showing a compound reverse-positive sedimentary rhythm. The highest resistance appears in the central part of the reservoir, over 27 Ωm. The dual-induction logging curve shows low water invasion and pure oil reservoir. The testing productivity reaches 15-30 t/d (Fig. 4-4).

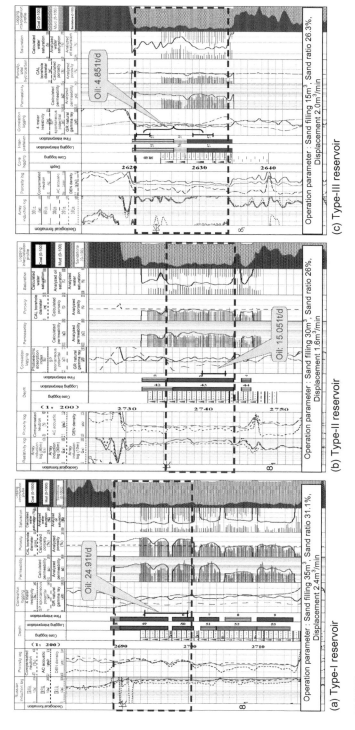

(a) Type-I reservoir

Operation parameter : Sand filling 35m³, Sand ratio 31.1%,
Displacement 2.4m³/min

Oil: 24.91t/d

(b) Type-II reservoir

Operation parameter : Sand filling 30m³, Sand ratio 26%,
Displacement 1.6m³/min

Oil: 15.051t/d

(c) Type-III reservoir

Operation parameter : Sand filling 15m³, Sand ratio 26.3%,
Displacement 2.0m³/min

Oil: 4.851t/d

FIGURE 4-3   The relationship between differences of three porosity curves and well-testing productivity.

Multisection composite reservoirs. The single-layer thickness in these reservoirs is usually small (<4 m). The barriers between oil layers are often over 10 m thick. With a resistivity of more than 27 Ωm, such reservoirs contain pure oil, with a single-layer testing productivity between 6 and 10 t/d and a total of 8~20 t/d (Fig. 4-5).

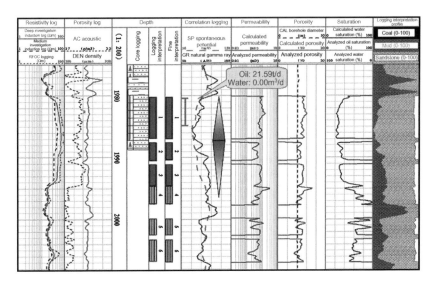

FIGURE 4-4   Log interpretation results from the Chang-4 + 5 reservoir in well Yuan-48.

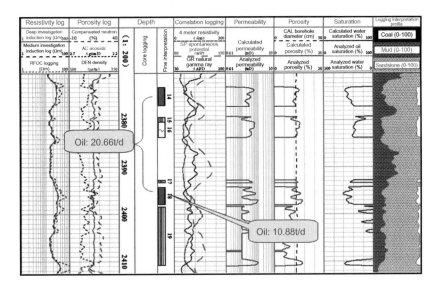

FIGURE 4-5   Log interpretation results from the Chang-6 reservoir in well Geng-27.

Oil-above-water reservoirs. These reservoirs are often contained in thick compound sandbodies, whose resistivity is more than 16 Ωm on the top, where oil and water coexist, while at the bottom there are only water layers. Well testing produces both water and oil, with a productivity of $4 \sim 10$ t/d, more water than oil (Fig. 4-6).

Oil-water layers. The single sandbody that contains such layers is often thin and vertically heterogeneous, with many interlayers developed. Single-layer interpretation shows they are oil-water reservoirs, with a resistivity between 16 and 27 Ωm. Well testing produces both water and oil, with a productivity of $4 \sim 12$ t/d (Fig. 4-7).

The classification above shows that the testing productivity from massive reservoirs generally lies between 15 and 30 t/d; that in the multisection composite reservoirs the single-layer testing productivity ranges from 6 to 10 t/d and the total productivity from 8 to 20 t/d; that the oil-above-water reservoirs produce both water and oil, with a testing productivity of $4 \sim 10$ t/d, more water than oil; and that the dispersed oil-water layers also produce both water and oil, with a testing productivity of $4 \sim 12$ t/d. It is clear that the reservoir type is somewhat related to its productivity (Fig. 4-8).

The Productivity Index Method While establishing a logging prediction model of reservoir productivity, we should choose enough logging data closely related to reservoir productivity and other geological and drilling data, considering all the possible factors that may influence the reservoir quality and productivity such as lithology, reservoir type, porosity, permeability, saturation and effective thickness, and analyzing their effects on

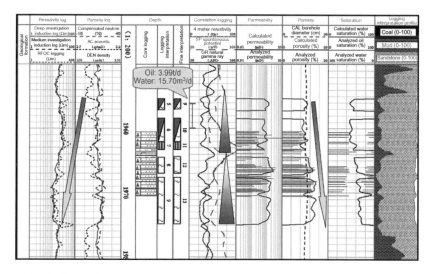

FIGURE 4-6   Log interpretation results from the Chang-4 + 5 reservoir in well Yuan-86.

reservoir productivity. Thus, from the data that reflect these factors, we can give them corresponding weights according to their own qualities and contributions to the reservoir quality and productivity, calculate through some model the reservoir evaluation indexes that reflect the weights, and then develop the indexes into a logging prediction model of reservoir productivity with a certain mathematical method.

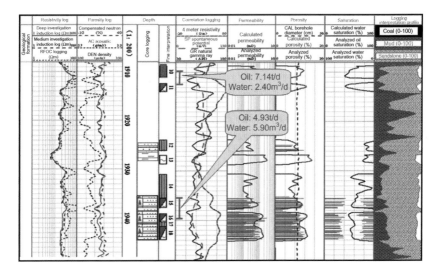

FIGURE 4-7   Log interpretation results from the Chang-4 + 5 reservoirs in well Yuan-106.

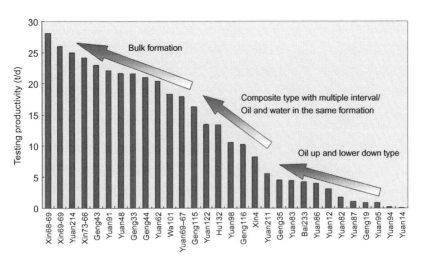

FIGURE 4-8   Relations between reservoir structure and testing productivity in the Chang-4 + 5 reservoir of Jiyuan.

The Composite Index Method Based on the Parameters of Physical Properties, Electric Properties, and Oil Saturation Studies have shown that no single reservoir parameter or logging parameter alone can reliably reflect reservoir productivity. In other words, one single parameter alone cannot be used to predict reservoir productivity. Considering that reservoir productivity is mainly affected by effective thickness, porosity, permeability, and oil saturation, we construct a composite index F to calculate the reservoir productivity to take all these parameters into consideration, as shown in the following:

$$F_{(R_t)} = H \times \phi \times k \times R_t$$

$$F_{(So)} = H \times \phi \times k \times So$$

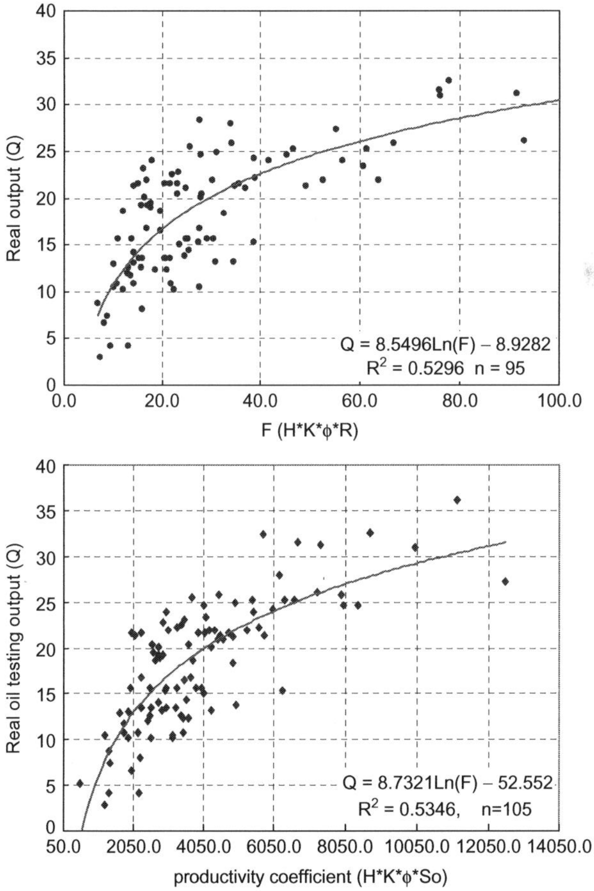

FIGURE 4-9   The crossplot of composite index and reservoir productivity.

The productivity prediction model through the composite index method:

$$Q = 8.5496Ln\ (F_{Rt}) - 8.9282\ R = 0.726,\ n = 95$$

$$Q = 8.7320Ln(F_{So}) - 52.552\ R = 0.730,\ n = 95$$

This model through the composite index method greatly improves the correlation between the influence factors and the productivity, catering to quick evaluation of reservoirs (Fig. 4-9).

The Method of Weighted Storativity Based on Reservoir Classification Our studies use as an example the Chang-8 reservoir in Zhenbei, which is characterized by complex pore structures and strong heterogeneity, with the whole sandbody bearing oil. Thus on the basis of fine reservoir classification, we apply the method of weighted storativity to predict productivity.

The thought underlying this method is as follows.

In the log evaluation of productivity, we use the classification principles discussed before to classify and grade the target intervals according to their contributions to productivity and determine the weighted storativity of a single contributor (Φeo.H, reflecting the oil storage capacity of the reservoir), so that we can determine the contributions (weights) of reservoirs of different types and grades and the "accumulated" productivity of the target interval by using the weighting method.

In the process of evaluation, priorities are given to the following three key parameters:

- The reservoir type of each contributing layer
- The storativity of a single layer (Φeo.H)
- The contribution rate (weight) of each unit layer

Based on these thoughts, the reservoir productivity can be expressed as

$$Q = \sum_{j=1}^{3} A_j \times I_j$$

where

Q—reservoir productivity

I_j—storativity of reservoirs of different types

A_j—contribution rates of reservoirs of different types to the storativity

The storativity I of reservoirs of different types can be expressed as

$$I = \sum_{i=1}^{n} \Phi_{eoi} \times H_i$$

in which Φeoi is the product of porosity and oil saturation of intervals of different types in the reservoir, H_i is the effective thickness of intervals of different types in the reservoir, and n is the number of intervals of different types in the reservoir.

Equipped with these thoughts and methods, we analyze the well testing data from 23 wells in the Chang-8 reservoir of Zhenbei, and then, by using three-variable statistical regression, identify the contribution rates of reservoirs of different types to storativity, i.e., 44.57 for Type-I, 14.33 for Type-II, and 6.55 for Type-III.

A logging productivity prediction equation is then established as follows:

$$Q = 44.57 \times I_1 + 14.33 \times I_2 + 6.55 \times I_3, \ R = 0.97, \ N = 23$$

The big differences among porosities, permeabilities, and storativities in reservoirs of different types make different contributions to productivity. Better reservior properties and higher oil saturations will contribute more to productivity. By analyzing the relationship between well-testing productivity and reservoirs of different types we get normalized coefficient weights of reservoirs of different types through the normalization method, i.e., 0.65 from Type-I, 0.25 from Type-II, and 0.1 from Type-III. Then we can establish the productivity calculation equation as

$$Q = 66.11 \times (0.65 \times I_1 + 0.25 \times I_2 + 0.1 \times I_3)$$

Figure 4-10 is a crossplot of predicted daily output calculated through this equation and real testing daily output, which shows a good correlation between the former and the latter, each distributed on one side of the line 45°, presenting a high accuracy.

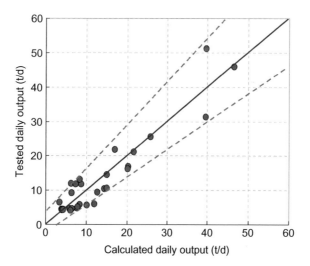

FIGURE 4-10   A crossplot of calculated daily productivity and real testing daily productivity of the Chang-8 reservoirs in Zhenbei.

The productivity prediction model established with weighted storativity in the Chang-8 reservoirs of Jiyuan and the Chang-6 reservoirs of Huaqing exhibits high accuracy and therefore, has good applicability.

The productivity prediction model for the Chang-8 reservoirs in Jiyuan:

$$Q = 45.74 \times (0.52 \times I1 + 0.28 \times I2 + 0.2 \times I3), \ R = 0.88, \ n = 29$$

The productivity prediction model for the Chang-6 reservoirs in Huaqing:

$$Q = 40.87 \times (0.46 \times I1 + 0.3 \times I2 + 0.24 \times I3), \ R = 0.93, \ n = 25$$

Applications of Quick Evaluation of Productivity

Log Interpretation Results from Exploration Wells and Appraisal Wells Fine interpretation with the quick evaluation method applied to key exploration wells and appraisal wells in 2009 shows that 20 wells are oil or oil-water layers, 19 of which produce commercial oil flows. Thus this method improves the accuracy of primary interpretation.

Optimization of Productive Zones The method of quick evaluation applied to regional productivity analysis helped select a favorable zone of 1.4-million-ton reserves in the Chang-8 reservoirs of Jiyuan and the block of Bai-153 in the Chang-6 reservoirs of Huaqing. It turns out that the real performance of these zones in the next year agreed quite well with the predicted data. Meanwhile, the evaluation of the Huang-3 well zone showed that the area from Yuan-26–96 to Yuan-31–95, unable to achieve the minimum profit, was not worth economic development, so the drilling of 22 wells in this area were postponed. In the area from the Bai-215 well to the Bai-301 well, under similar conditions, 29 wells were postponed.

In short, enhanced early evaluation means extension of exploration, which is of benefit to fast, scale and efficient development. Identifying the productive zones by combining pilot exploration, quick evaluation, and drilling can help reduce the construction of redundant facilities, promote scale production, and shorten the cycle of field construction. Such an overall strategy may promote the quick and efficient development of ultralow-permeability reservoirs and thus bring greater rewards to investment.

4.2 AUXILIARY EQUIPMENT FOR ADVANCED WATER INJECTION

4.2.1 Small, Closed Mobile Water Injection Skid

When a normal longstanding injection system is unfit for some dispersed blocks or areas of unproven reserves, the use of a small, closed mobile water injection skid (injection skid for short) can facilitate advanced injection (Fig. 4-11).

Such a skid is installed in the open air on the well site, consisting of a water tank, an injection pump, a set of water treatment devices, a control system, valves, pipelines, measuring meters, and the skid base, and integrated with water supply, filtering, dosing, pressure boost, measuring, and backflow. All the devices, valves, and pipes are installed on the 8.2 m × 2.4 m skid base. The mobile skid plays an important role in the development of ultralow-permeability reservoirs through advanced injection because it not only conveys the design concept of digital management, standardized design, and modular construction, but it also has the functional advantages of short procedures, easy transport, and convenient operation.

Technological Procedures

After flowing into the tank, the water is treated with closed buffering, settlement, pump feeding, and fine filtering for pressure boosting through the injection pump. The qualified water thus obtained is then measured and adjusted by the trunk injection lines before being transmitted out towards the injection pipe network for water allocation.

Supporting Facilities

The mobile injection skid installed in the open air has an electric charge of 100 kW and needs an exclusive pole-type substation on the well site, containing a transformer of S10-M-160/10 160 kVA. The power source of the frequency conversion instrument cabinet is connected to a low-voltage distribution box in the new substation.

The injection skid and the water supply well, which are designed at the same time with centralized control and monitoring, can be operated interactively. A new well is drilled within 200 m from the skid to supply water. The type of borehole pump for the water well depends on its completion

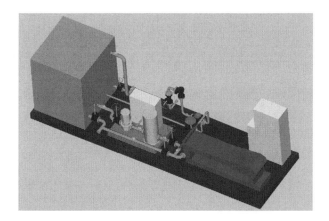

FIGURE 4-11   Small, closed mobile water injection skid.

parameters on the site, with a rated flow of 15 m^3/h. An electromagnetic flowmeter and a digital pressure transmitter are installed at the wellhead, which will send pressure signals to the instrument cabinet for it to control the frequency of the borehole pump according to its outlet pressure changes so that its outlet flow can be maintained at 15 m^3/h, which will ensure that the pump operates smoothly.

4.2.2 The Unattended Smart Flow-Regulating Valve Complex

This kind of complex is installed on the dendritic trunk pipelines for wellhead water distribution (Fig. 4-12), thus abolishing the conventional water-distributing stations and realizing the first-order station arrangement, which is

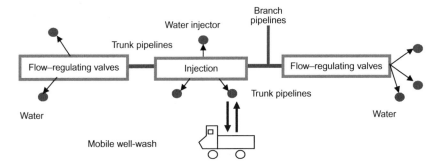

FIGURE 4-12   Dendritic trunk pipeline of flow-regulating valve complex.

FIGURE 4-13   The smart flow self-controller.

especially fit for the cluster-well sites and the characteristics of losses plateau and can save 43.6% of the pipelines for water injection.

The smart self-controller (Fig. 4-13) attached to the valve complex integrates flow measurement with its adjustment. Characterized by a compact structure, high accuracy, and efficient flow stabilization, the unattended controller can not only adjust the flow rate in a closed cycle, but also collect and monitor the key production data automatically.

4.2.3 The Integrated Digital Skid Pressurizer

This pressurizer, combining a knockout surge drum, a heating furnace, a gas-liquid separator, and effluent pumps on a skid, and furnished with a sophisticated smart control system, can practice remote control for various methods of oil transmission, such as heated pressurization, pressurization without heating, heated separate buffering pressurization, and heating without pressurization (Fig. 4-14). With the advantages of integrated functions, fast installment, easy handling, energy saving, environmental friendliness, convenient transport, and repeated use, the device is appropriate for quick rolling development. After its application in Changqing, this unattended device optimized procedures with good economic and social returns. The successful development of the device fills in the gaps in the field of integrated technology for petroleum gathering

FIGURE 4-14   The integrated digital skid pressurizer.

and transmission in China, lays a foundation for practice of oilfield development through advanced injection, and thus is an important innovation in oilfield ground engineering.

With reference to the situation in the Changqing Oilfield, we have designed and made three series of integrated digital skid pressurizers, with a respective daily oil output and pressure of 240 m^3, 2.4 MPa; 240 m^3, 3.2 MPa; and 120 m^3, 2.4 MPa. Through collecting and analyzing field data, we will further optimize and update all procedures and techniques in equipment design and manufacture to improve the matching between the output to be handled and the pressure, the corrosion resistance of the chimney and the flaring system, the gas-entraining capacity in multiphase flow transportation, the overall safety and reliability, instrument selection, and tubing procedures.

Constant innovation and improvement of the auxiliary technologies makes it possible for advanced injection to be widely used in Changqing. The application of a mobile injection skid, smart flow-regulating valve complex, and integrated digital skid pressurizer helps increase the scale of advanced injection year by year. By 2009, the output harvested through advanced injection had accounted for over 90% of the total in the whole Changqing Oilfield Company, with the per-well output increasing by 20−30% on average.

4.3 NODAL OPERATION OF ADVANCED WATER INJECTION

Technology is closely related to management. Quick evaluation of reservoirs and their productivity helps identify favorable zones for oil and gas production. The application of the mobile injection skid, smart flow-regulating valve complex, and integrated digital skid pressurizer helps solve the technical problems in the implementation of surface projects. An overall, systematic, and effective control of advanced injection must rely on updated management. Nodal operation of advanced injection and overall planning of a surface framework prove an effective method to facilitate the use of technologies and procedures for advanced injection.

4.3.1 Guidelines for Advanced Injection

Advanced injection should observe the following guidelines:

- Fit for building productivity.
- The whole system should be rational and matching.
- Technologies should be updated and suitable.
- Resources should be comprehensively utilized.
- Safe, environmental friendly, and energy saving.
- Convenient and easy management.
- Investment and control in place.

4.3.2 Integrated Planning

We adhere to the principle of integrated exploration and development, the principle of "three-into-one" optimization of the geological, development, and surface systems, and the principle of "integral deployment, overall planning, accelerated implementation and priority to the framework," so as to put the surface framework system in place at a stroke and carry out water injection in advance.

A high emphasis is placed on the principles of "standardized design, modularized construction, digital management and market-oriented operation." The surface construction observes the principles of "two adapted," "two enhanced," and "two reduced." "Two adapted" means that it should be adapted to large-scale construction and rolling development. "Two enhanced" means that it should enhance the production efficiency and the construction quality, and "two reduced" means that the surface construction should reduce security risks and total costs.

Keeping in mind the strategy of low-cost development, we promote innovative surface techniques and optimize surface construction modes to provide good conditions for flat management. From 2009 to 2011, we built 18 union stations, 43 water injection stations, 341 water wells, and 8 substations, at a cost of 2.968 billion yuan. For example, 4 union stations and 22 water injection stations were built in 2009, ahead of schedule.

4.3.3 Nodal Operation of Advanced Injection

Surface construction is the premise of productivity building. Building surface pipes and stations in advance may facilitate the implementation of advanced injection. The goal of nodal operation of surface construction for advanced injection is "36911":

- "3"—the project design and other preliminary work for the surface construction to be finished by the end of March
- "6"—60% of the stations to be put into operation by the end of June
- "9"—9% of the stations to be put into operation by the end of September
- "11"—all surface construction projects to be completed and put into operation by the end of November

From what has been discussed, it is easy to see that the practice of advanced injection is not only a technical job but also a systematic program. Quick evaluation of reservoirs and their productivity helps identify favorable petroliferous zones, so as to minimize the risks of drilling low-PI wells and dry wells, provide a guide to the development strategy, and improve development efficiency. While constructing the ground system in advance can facilitate advanced injection, nodal control introduces updated management into its process. Combined with a systematic planning of the surface construction framework, the application of a mobile water injection skid, smart flow-regulating valve complex, and integrated digital skid pressurizer helps increase the scale of advanced injection year by year, which greatly contributes to the speedy and efficient development of the ultralow-permeability lithological reservoirs in the Ordos Basin.

Practice of Advanced Water Injection

Initiated in 2001, advanced water injection, as it kept improving, was adopted on an increasingly larger scale in Changqing (Fig. 5-1), especially in the reservoirs of Chang-8, Chang-6, and Chang-4 + 5 in Xifeng, Jing'an, Nanliang, Ansai, and Jiyuan oilfields in 2001−2009, covering an oil-bearing area of 1522 km^2, geological reserves of 8.2×10^8 t, and a productivity of 1344×10^4 t. It achieved three good results, as presented below.

Considerably increasing the per-well output of ultralow-permeability Triassic reservoirs. Compared with conventional synchronous or delayed flooding, advanced injection increased the average per-well output by 20 to 30%. Statistics from the 1061 advanced-flooded producers show that they have an average per-well output of 4.39 t/d at the initial stage, 0.63 t/d higher than adjacent nonadvanced-flooded wells (Table 5-1).

Slowing down the output decline in ultralow-permeability Triassic reservoirs. In our practice of overall advanced injection in the development of the Xifeng Oilfield, a higher rate of the capacity-building producers reach our designated productivity, with less decline than those in the same reservoirs

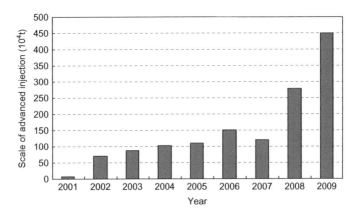

FIGURE 5-1 Increasing scale of advanced injection in Changqing, 2001−2009.

TABLE 5-1 A Comparison between the Effects of Advanced and Nonadvanced Injection

Block	Production time	Horizon	Advanced-flooded wells					Nonadvanced-flooded wells				
			Well number	QI (m³)	Qo (t)	fw (%)	PFL (m)	Well number	QI (m³)	Qo (t)	fw (%)	PFL (m)
Wangyao	2001	Chang-6	21	5.48	3.00	35.6	1387	14	5.08	2.10	51.4	1412
Yinhe	2001	Chang-6	14	6.10	4.50	13.2	1433	18	4.55	3.10	19.8	1467
West Nanliang	2003–2004	Chang-4 + 5	72	5.52	3.97	15.4	1518	31	5.47	3.62	22.1	1597
Baima	2002–2004	Chang-8	298	8.61	6.92	5.4	1178	49	7.65	6.20	4.7	1281
Dongzhi	2003–2004	Chang-8	27	6.47	4.47	18.7	1157	17	5.65	3.97	17.3	1531
Panguliang	2002–2004	Chang-6	92	7.96	5.32	21.4	1292	42	7.52	4.94	22.7	1468
Dalugou	2002–2004	Chang-6	195	8.49	5.17	28.4	1247	19	6.51	3.54	36.0	1353
Baiyushan	2002–2004	Chang-4 + 5	133	8.07	4.83	29.6	1160	27	6.92	3.98	32.3	1165
Wu-420	2005–2006	Chang-6	35	6.90	4.50	34.8	1650	120	4.20	3.80	9.5	1700
Bai-209	2006	Chang-6	44	5.28	4.62	12.5	1399	72	4.66	3.98	14.7	1488
Bai-157	2007	Chang-4 + 5	17	5.88	2.83	50.6	1320	86	5.34	3.15	40.9	1343
Bai-168	2008	Chang-8	32	3.75	3.51	6.3	1659	51	3.58	3.21	10.3	1613
Luo-1	2008–2009	Chang-8	81	4.19	3.45	16.8	1596	260	4.12	3.35	18.2	1669
Total			1061	6.36	4.39	31.0	1384	806	5.48	3.76	31.3	1468

that do not use advanced injection. Statistics show that 75% of the producers that drilled in 2003−2008 in the Baima block of Xifeng reached our designated productivity, while 82% of those drilled in these years in the Triassic reservoirs in other blocks did so (Fig. 5-2). The total decline rate of producers drilled since 2004 in the Baima block was below 10% in the third year of production, while the overall decline of producers in all blocks in the Triassic reservoirs reached 14.2% (Fig. 5-3) in the past 4 years.

Building an effective displacing pressure system, thus improving the efficiency of production in the ultralow-permeability Triassic reservoirs. Advanced injection produced responses in as high as 71.7% of oil wells on average, as shown in Table 5-2.

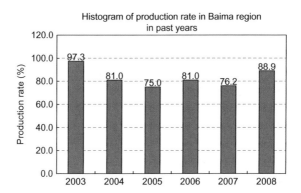

FIGURE 5-2 A comparison of percentages of capacity-building producers whose output reached the designated objectives in Baima, 2003−2008.

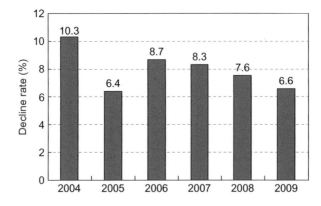

FIGURE 5-3 A comparison of output decline rates of the capacity-building producers in Baima, 2004−2009.

TABLE 5-2 Responses of Producers to Advanced Injection

Block	Production time	Horizon	Response rate (%)
Yinhe	2001	Chang-6	78.6
West Nanliang	2003–2004	Chang-4 + 5	75.6
Baima	2002–2004	Chang-8	91.1
Dongzhi	2003–2004	Chang-8	68.5
Panguliang	2002–2004	Chang-6	89.2
Dalugou	2002–2004	Chang-6	62.4
Baiyushan	2002–2004	Chang-4 + 5	65.3
Wu-420	2005–2006	Chang-6	75.3
Bai-209	2006	Chang-6	64.1
Bai-157	2007	Chang-4 + 5	63.8
Bai-168	2008	Chang-8	57.2
Luo-1	2008–2009	Chang-8	68.7
Average			71.7

5.1 DEVELOPMENT OF THE JING'AN OILFIELD

5.1.1 General Overview

Exploration of the oilfield began in 1983. The year 1997 saw the beginning of industrial production focused on the Chang-6 Triassic reservoirs, which have an average buried depth of 1890 m, an initial pressure of 12.2 MPa, a pressure coefficient of 0.67, and a saturation pressure of 7.02 MPa. The crude in place has a viscosity of 2.0 mPa s and a density of 0.767 g/cm^3. In addition, the reservoirs have a temperature of 54.39 °C, an average effective thickness of 14.9 m, a porosity of 12.8%, and a permeability of 1.81×10^{-3} μm^2.

At the end of 2007, the oilfield had 3533 producers, with 3252 at work, and 1194 injectors, with 1132 in operation. The average per-well output was 4.8 m^3 of fluids a day, including 2.7 t oil. The composite water-cut reached 49.1% in the producers. The average water injection rate per well was 29 m^3/d, at a monthly injection-to-production ratio of 1.54 and a cumulative ratio (cumulative voidage replacement ratio) of 1.26. The total production in 2007 reached 331×10^4 t, at a producing rate of 1.07% of the geologic reserves and with a recovery of 6.91%.

5.1.2 Technological Policies

With reference to theoretical studies, we practiced in the Jing'an Oilfield the technological policy of water injection six months in advance, with an intake per unit thickness of 1.8 to 2.3 m^3/d m. The injection timing schemes of the oilfield underwent three phases: delayed injection (more than half year after commissioning), synchronous injection, and advanced injection, producing different results. Practice shows that a reasonable technological policy at the initial stage of production enables a well to have a high productivity, which may decline more slowly and remain stable for a longer period of time.

The Best Time to Begin Water Injection Is Six Months before Commissioning

Our experience shows that a longer period of advanced injection may lead to a higher producing energy and initial output, with less decline. On the contrary, an insufficient lead time will have little effect on reducing the output decline, or increasing the per-well output, because it will not make up for enough producing energy. According to the statistics of the three typical advanced-flooded well groups of Jing'an (Table 5-3), a producer with ten months of advanced flooding has an initial output of 14.5 t/d, which declines to 9.7 t/d six months after commissioning. At the same time, the initial per-well output from adjacent producers with nonadvanced flooding is 9.04 t/d, which rises to 9.05 t/d six months after commissioning. For the producers with eight months of advanced injection, their per-well output is 9.37 t/d at the beginning and declines to 6.78 t/d six months after commissioning, and that of the adjacent wells with nonadvanced flooding is 13.1 t/d, which declines to 4.25 t/d six months after commissioning. For the producers with three months of advanced flooding, their per-well output is 7.59 t/d at the beginning and declines to 4.03 t/d six months after commissioning, and that of the adjacent wells with nonadvanced flooding is 9.8 t/d, which declines to 4.7 t/d six months after commissioning. It can be seen from these statistics that all producers with advanced flooding have a higher initial output, and the longer the advanced flooding, the higher the output of a flooded producer six months after commissioning as well as at the initial stage. However, a longer injection time may involve higher investment. All factors considered, we conclude that the proper timing for water injection is three to six months before commissioning.

Selection of Reasonable Injection Parameters

In terms of the relationship of the injection intensity to the per-well output and water-cut, six to nine months of advanced injection in the Hulangmao Oilfield, with an initial volume of 28 to 35 m^3/d and an intensity of 1.8 to 2.3 m^3/d m, produced a high initial oil output and a low composite water-cut

TABLE 5-3 Output Statistics of Producers with Advanced Injection of Different Timing Schemes in Wuliwan Block 1

Injector number	Lead time (months)	Corresponding flooded producers							Adjacent producers with nonadvanced flooding						
		Well number	Test daily output		Initial daily output		Current daily output		Well number	Test daily output		Initial daily output		Current daily output	
			Oil (t)	Water (t)	Oil (t)	Water (t)	Oil (t)	Water (t)		Oil (t)	Water (t)	Oil (t)	Water (t)	Oil (t)	Water (t)
Liu-90-46	10	Liu-90-47	36.6	0	14.2	0.7	5.9	0.5	Liu-91-45	18.7	0	13.5	0.2	6.7	0.1
	10	Liu-91-46	34.2	0	14.6	1.2	13.5	2.2	Liu-90-45	19.5	0	11.6	0.8	8.5	0.1
									Liu-89-45	28.3	0	9.1	0.1	7.9	0.2
									Liu-89-46	33.2	0	1.9	1.8	13.1	0.7
Average	10		35.4	0	14.4	0.9	9.7	1.4		24.9	0	9.0	0.7	9.1	0.3
Liu-131	8	Liu-84-35	24.0	0	9.2	0.9	4.7	0.0	Liu-83-34	20.1	0	16.4	1.2	2.1	0.0
	8	Liu-85-35	25.0	0	9.6	0.8	8.0	0.1	Liu-83-35	22.6	0	14.6	1.2	4.8	0.0
	7	Liu-85-36	25.5	0	10.0	1.2	8.7	0.1	Liu-84-33	16.3	5.6	10.0	1.0	4.4	0.0
	8	Liu-85-34	22.8	0	8.7	0.9	5.7	0.1	Liu-83-33	16.8	11.1	11.5	0.6	5.7	0.1
Average	7.75		24.3	0	9.4	0.9	6.8	0.1		19.0	4.2	13.1	1.0	4.3	0.0
Liu-78-32	3	Liu-77-31	19.2	0	8.2	0.3	2.0	0.0	Liu-78-31	21.9	0.0	9.5	0.5	5.5	0.1
	5	Liu-77-32	18.3	0	6.1	0.5	2.6	0.0	Liu-79-33	21.3	7.5	10.3	0.5	3.9	0.1

(Continued)

TABLE 5-3 Output Statistics of Producers with Advanced Injection of Different Timing Schemes in Wuliwan Block 1 (cont.)

Injector number	Lead time (months)	Corresponding flooded producers							Adjacent producers with nonadvanced flooding						
		Well number	Test daily output		Initial daily output		Current daily output		Well number	Test daily output		Initial daily output		Current daily output	
			Oil (t)	Water (t)	Oil (t)	Water (t)	Oil (t)	Water (t)		Oil (t)	Water (t)	Oil (t)	Water (t)	Oil (t)	Water (t)
	6	Liu-77-33	21.6	0	5.7	0.3	5.5	0.2							
	3	Liu-78-33	19.5	0	8.6	1.3	4.8	0.1							
	3	Liu-79-31	13.5	0	9.1	0.6	5.3	0.0							
	3	Liu-79-32	16.2	0	8.0	0.3	4.0	0.0							
Average	3		18.1	0	7.6	0.5	4.0	0.1		21.6	3.8	9.9	0.5	4.7	0.1

after well response. With advanced injection for three months at a rate of about 25 m³/d and an intensity of 1.5 to 1.8 m³/d m, the initial output increased to a relatively small degree, while the composite water-cut rose fast (Figs. 5-4 and 5-5). Thus based on the experience gained in Hulangmao and Wuliwan, we chose an initial injection-to-production ratio of 1.2, and to achieve a per-well productivity of 5.5 t/d, we decided on an injection rate

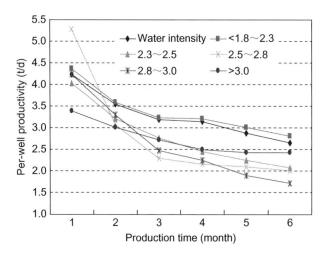

FIGURE 5-4   The relationship between advanced-injection intensity and per-well productivity in Hulangmao.

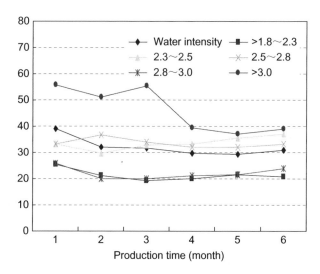

FIGURE 5-5   The relationship between advanced-injection intensity and composite water-cut in Hulangmao.

between 30 and 35 m^3/d and a water intake between 1.8 and 2.3 m^3/d m per unit thickness, which was adjusted with the progress of production.

5.1.3 The Effects

5.1.3.1 Advanced Injection Increases Reservoir Pressure Considerably, Favorable for Establishing an Effective Displacing Pressure System

It can be seen from Table 5-4 that the reservoir pressure of well Liu-79-40 with a long injection time reached as high as 119% of the original pressure, while in well Liu-77-59 with a delayed injection, the reservoir pressure was only 91.8% of the original. It is obvious that advanced injection can supply energy for the reservoir. The longer the injection time, the higher the pressure will be maintained.

Statistics of the five producers in the advanced-flooded well groups in Wuliwan show that the reservoir pressure is kept between 10.5 and 13.5 MPa, that of the four producers in the synchronous-flooded well groups remains between 9.5 and 10.0 MPa, while that of the four producers in the delayed-flooded well groups is only 8.6 to 9.2 MPa (with the reservoir-choked wells excluded). This means that an earlier start of water injection helps keep the reservoir energy at a higher level.

TABLE 5-4 Statistics of Reservoir Pressure with Advanced Injection in Wuliwan Block 1

Well zone	Well number	Time of measurement	Pressure (Mpa)	Percent of the initial (%)	Notes
ZJ-53	Liu-77-59	2001.6.14	11.25	91.8	Production before injection
	Liu-79-60	2001.9.8	12.4	101	9 months of advanced injection in wells 80-60
Total			23.65		
Average			11.83	96.6	
Liu 78-40 Liu 80-40	Liu-79-40	2001.8.20	14.56	119	6 months of advanced injection in wells 78-40; 44 months of advanced injection in wells 80-40

Note: The initial pressure is 12.25 MPa.

5.1.3.2 Advanced Injection Leads to a High Initial Daily Oil Output, Which Stabilizes for a Longer Period of Time and Declines More Slowly

In the Wuliwan block of Jing'an, advanced injection was initiated in some skeleton wells in 1997. Statistics show that the 30 advanced-flooded producers enjoyed a stable high output over 6.0 t/d for a long time, and the synchronous-flooded 165 producers maintained an output of 5.0 t/d after two years of production. In the 160 delayed-flooded wells, the initial per-well output declined dramatically and for a longer time, with very slow pressure buildup. Even after a producer responded, its output still increased very little (by 0.9 t/d per well on average), and stayed at about 4.0 t/d (Fig. 5-6).

A comparison between the result of advanced injection and that of synchronous injection in the Chang-6 reservoir of the Wuliwan block shows that, under similar conditions of reservoir properties, reservoir stimulation, and well patterns, the advanced-flooded producers have a higher initial daily output, which stabilizes longer and declines more slowly than the synchronous-flooded ones. (Table 5-5 and Fig. 5-7).

In the Baiyushan block of Jing'an, advanced injection supplemented the reservoir energy effectively. The point-pressure tests in 40 producers showed that the average pressure was 11.12 MPa, 106% of the initial value. In addition, the output decline slowed down, with the liquid output, oil output, water-cut, and producing fluid level all staying steady (Fig. 5-8).

In 2003, when advanced injection was not carried out in the Chang-6 reservoir in the Dalugou block of Jing'an, a producer had an average per-well output of 3.59 t/d in the first three months of production, with a water-cut of 30.5%. In 2004, when advanced injection was practiced across the whole block, the average per-well oil output in the first three months was raised to 6.15 t/d, with the water-cut lowered to 20.6% (Table 5-6). It is obvious that

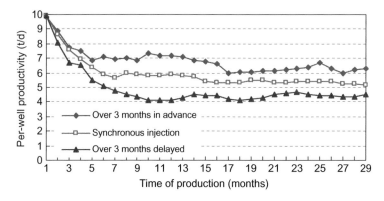

FIGURE 5-6 Results of different waterflood-timing schemes in the Wuliwan block, Jing'an.

TABLE 5-5 Reservoir Properties and Oil-Testing Results from the Typical Producers in the Wuliwan Block

Well group	Well name	Commissioning time	Producer type	Water injection timing	Reservoir	Effective thickness
Liu-80-34	Liu-81-34	1998.4	Edge well	4 months in advance	Chang-6	11.6
	Liu-79-34	1997.12	Edge well	Synchronous	Chang-6	13.4
Liu-131	Liu-85-35	1998.7	Corner well	8 months in advance	Chang-6	20.6
	Liu-83-34	1997.8	Corner well	4 months delayed	Chang-6	14.4
Liu-78-34	Liu-78-33	1998.8	Edge well	8 months in advance	Chang-6	11.4
	Liu-79-34	1997.12	Edge well	Synchronous	Chang-6	13.4

Well group	Well name	Properties		Stimulation		Oil testing	
		K (10^{-3} μm^2)	Φ (%)	Type	Propants (m^3)	Daily oil output (t)	Daily water output (m^3)
Liu-80-34	Liu-81-34	4.5	12.1	Gel	20	20.7	0
	Liu-79-34	4.83	12.8	Gel	21	13.3	0
Liu-131	Liu-85-35	3.57	11.95	Gel	27	25.2	0
	Liu-83-34	7.02	11.67	Gel	21	18.7	0
Liu-78-34	Liu-78-33	3.34	12.7	Gel	24	18	0
	Liu-79-34	4.83	12.8	Gel	21	13.3	0

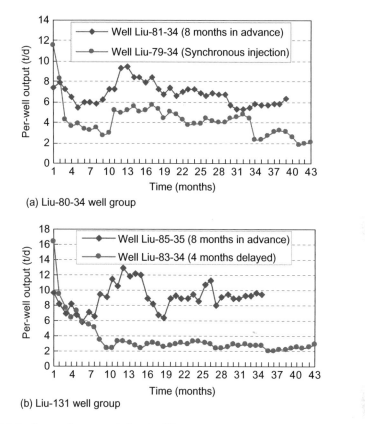

(a) Liu-80-34 well group

(b) Liu-131 well group

FIGURE 5-7 Curves of output variations in different water-timing schemes, Wuliwan block.

the liquid and oil outputs increased considerably while the water-cut declined significantly. An effective displacing pressure system was built (Fig. 5-9).

5.2 DEVELOPMENT OF THE ANSAI OILFIELD

5.2.1 General Overview

The Ansai Oilfield was founded in 1983 and three pilot well groups, i.e., Sai-1, Sai-5, and Sai-6, were drilled successively in 1985. Large-scale injection was implemented in 1990. Microfractures are well developed in the oilfield, as shown in one third of the cores from these wells.

The Chang-6 reservoirs in these wells have an average buried depth of 1130 m, an initial pressure from 8.31 to 10 MPa, a pressure coefficient from 0.7 to 0.9, and a saturation pressure from 4.65 to 6.79 MPa. The oil in these reservoirs has a viscosity of 4.85 to 7.86 mPa s on the ground, a viscosity of 1.96 to 7.8 mPa s, in place and a density of 0.85 g/cm^3. In addition, the

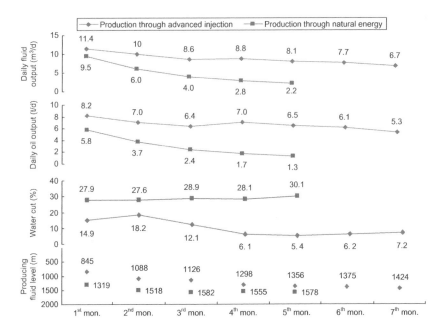

FIGURE 5-8 A comparison between the result of advanced injection and that of natural production in the Baiyushan block.

reservoirs in these wells have a temperature of 45 °C, an effective porosity from 11% to 15%, and a permeability of $1 \sim 3 \times 10^{-3} \ \mu m^2$.

At the end of 2007, the oilfield had 4281 production wells, with 3851 in operation, and 1539 water injectors, with 1417 at work. The average per-well liquid output was 3.6 m³/d and the average oil output was 1.8 t/d, with a composite water-cut of 49.63%. The average per-well injection rate was 24 m³/d, with a monthly injection-to-production ratio of 2.0 and a cumulative ratio of 1.74. In 2007, the oilfield produced a total of 255.08×10^4 t of oil, with a withdrawal rate of 0.92% of the total geologic reserves and a recovery factor of 7.85%.

5.2.2 Technological Policies

5.2.2.1 The Best Injection Timing

Theoretical studies show that the best injection timing in Wangyao is six months in advance. According to the actual performance (Fig. 5-10) and the effects of the different injection timing schemes (Table 5-7), the producers flooded for nine or six months in advance show little decline of initial output and a short response time, after which the oil output gets obviously higher than that of those flooded for three months in advance. On the other hand, a

TABLE 5-6 A Comparison of Outputs in a Typical Block of Dalugou

| Year | Well numbers | Electrical resistivity ($\Omega \cdot m$) | Porosity (%) | Permeability ($10^{-3}\ \mu m^2$) | Water saturation (%) | Average in the first three months | | | |
						Liquid output (m^3/d)	Oil output (t/d)	Water-cut (%)	Producing fluid level (m)
2003	42	17.39	12.58	3.19	57.88	6.15	3.59	30.5	1370
2004	82	17.65	12.68	3.22	53.55	9.3	6.15	20.6	1330

comparison between the producers for six months of advanced injection with those for nine months shows that their response periods and daily outputs in the response period and in the period of stable production are generally the same. In spite of the relatively higher initial output and lower ratio of the output in the response and stable periods to the initial output in the producers with nine months of advanced flooding, we conclude that six months before commissioning is the reasonable injection timing for the ultralow-permeability Chang-6 reservoirs in Ansai, taking the overall economic efficiency into consideration.

5.2.2.2 Selection of Reasonable Injection Parameters

Theoretical calculations indicate that the injection intensity should be kept at 2.5 m^3/d m. In the Wangyao block with six or nine months of advanced injection, the initial per-well rate was kept at 40 to 50 m^3/d, with an intensity of 2.5 m^3/d m, which brought a high output after the producers began to respond. By contrast, the producers with three months of advanced injection, at an initial per-well rate of 35 m^3/d and an injection intensity of

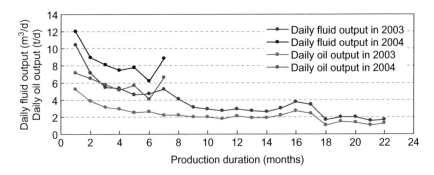

FIGURE 5-9   Curves of oil outputs with advanced injection in Dalugou in 2003 and 2004.

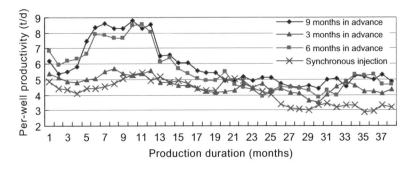

FIGURE 5-10   Results of different injection timing schemes in the Wangyao block.

TABLE 5-7 Results of Different Injection Timing Schemes in the Wangyao Block, Ansai Oilfield

Injection timing	Initial output (t/d)	Output before response (t/d)	Ratio to the initial (%)	Maximum output after response (t/d)	Ratio to the initial (%)	Stable output after response (t/d)	Ratio to the initial (%)	Response period (month)
9 months in advance	6.2	5.38	86.8	8.6	138.7	4.64	74.8	2 to 4
6 months in advance	6.8	5.92	87.1	8.5	125	4.63	68.1	2 to 4
3 months in advance	5.4	4.08	75.8	5.42	100.7	4.1	76.2	4 to 6

1.5 ~ 2.0 m³/d m, showed little increase in output after response (Figs. 5-11 and 5-12). This implies that the injection rate and intensity should both be increased before well response.

Meanwhile, our experience in Ansai shows that an injection intensity of less than 3.0 m³/d m may result in less intralayer fingering and thus avoid early water breakthrough. For this reason, we decided that the advanced injection intensity should be less than 3.0 m³/d m. In general, the initial per-well rate and the injection intensity should be kept at 40 to 50 m³/d and 2.5 to 3.0 m³/d m, respectively.

5.2.3 Results of Advanced Water Injection

With reference to the results from different injection timing schemes in the Chang-6 Triassic reservoirs, we carried out advanced injection tests in Ansai in 2001 for the purpose of maximizing the per-well productivity, slowing down the decline of initial output, and optimizing the relevant policies and technical parameters of advanced injection. The test involved 12 well groups (seven in southwest Wangyao and the other five in southwest Xinghe),

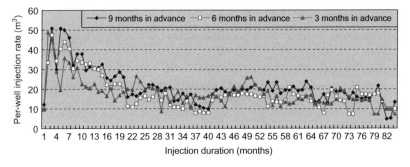

FIGURE 5-11   Curves of daily injection rates of different injection timing schemes in Wangyao.

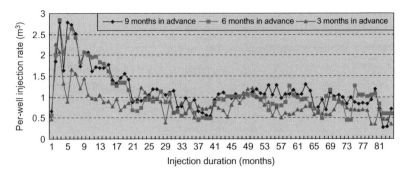

FIGURE 5-12   Curves of injection intensities of different injection timing schemes in Wangyao.

including 47 production wells. In the Wangyao groups, water injection lasted 33 to 132 days before commissioning, averaging 80 days, at a daily per-well rate of 41 m^3. In the Xinghe groups, water injection lasted 64 to 107 days before commissioning, averaging 83 days, at a daily per-well rate of 39 m^3 (Table 5-8).

5.2.3.1 Patterns of Pressure Changes

For knowledge of how well the reservoir pressure is maintained before commissioning, we followed up the pressure changes in 11 wells (six in southwest Wangyao and five in southwest Xinghe) through a fortnight pressure test of the monitored producers, totaling 53 well times (Table 5-9).

- There are three characteristics in the pressure variations.
- The rate of pressure rise tended to decline.
- In Wangyao, the rate of pressure increase for all six monitored wells started high and then declined. In the initial period (the first 40 days), the pressure rose at a rate of 0.2 to 0.67 MPa per month, averaging 0.35 MPa. Afterwards, the rise rate declined, staying at 0.02 to 0.08 MPa per month. For example, with an injection rate of 40 to 60 m^3 and an injection intensity of 1.89 to 2.86 m^3/d m into the Wang 34-015 injector, its flooded producers, namely Wang 34-014 and 35-016, experienced a pressure rise rate that declined from 0.67 to 0.02 MPa per month.
- The pressure values tended towards the same.
- From the six monitored wells in southwest Wangyao, two months of advanced water injection kept the producer pressures stable, tending towards the same value (between 9.60 and 9.88 MPa).
- There are fewer monitoring points in southwest Xinghe, with the pressure in most producers still rising. But according to the pressure changes in five monitored producers in this area, the pressures in different producers also tended towards the same (between 12.55 and 12.8 MPa) (Fig. 5-13).

TABLE 5-8 Physical Properties in Advanced-Flooded Production Wells

Block	Layer	Effective thickness (m)	Resistivity ($\Omega \cdot$ m)	Permeability (10^{-3} μm^2)	Porosity (%)	Water saturation (%)
Southwest Wangyao	Chang-6	22	16.9	1.666	13.1	54.7
Southwest Xinghe	Chang-6	18.2	23.8	1.42	11.7	52.7
Average	Chang-6	20.5	19.6	1.57	12.5	54

TABLE 5-9 Pressure Monitoring of Advanced-Flooded Producers

Well no.	Pressure value (MPa)						Average pressure rise rate (MPa/month)
	1st	2nd	3rd	4th	5th	6th	
Wang-29-106	9.21	9.37	9.53	9.57	9.59	9.6	0.14
Wang-30-018	9.38	9.43	9.47	9.51	9.51	9.52	0.05
Wang-31-017	9.39	9.58	9.78	9.96	9.87	9.88	0.18
Wang-31-018	9.22	9.32	9.44	9.55	9.57	9.58	0.13
Wang-34-014	8.84	9.22	9.61	9.65	9.65	9.66	0.3
Wang-35-016	9.06	9.42	9.81	9.83	9.83	9.84	0.28
Xing-6-01	12.24	12.48	12.55	12.49			0.15
Xing-7-004	10.36	10.69	10.91	11.22			0.53
Xing-5-002	11.61	12.34	12.72				0.95
Xing-7-04	10.83	11.26	11.63				0.69
Xing-7-05	10.98	11.36	11.63				0.56

(a)

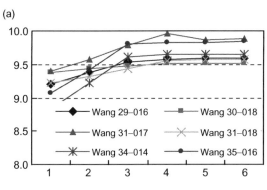

(b)

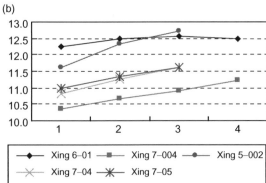

FIGURE 5-13 Pressure-monitoring curves in the Ansai Oilfield.

The Rate of Pressure Rise Was Relevant to the Time Interval between Oil Testing and Pressure Measuring

An analysis of the factors that influence the pressure rise shows that pressure measurement in the 11 wells lasted 7 to 32 days after oil testing. In this period, the shorter the time interval from oil testing to pressure measuring, the higher the pressure rise rate at the initial stage, and vice versa (Fig. 5-14).

In addition, from the 11 monitored producers and the injection time of their corresponding injectors, the average injection time was only nine days before the pressure measurement, thus the injected water had little influence on the pressure rise.

It can then be concluded that the ultralow-permeability Chang-6 reservoirs in Ansai are still at the elastic recovery stage two to three months after oil testing, when the well pressure gradually moves towards the initial value, and that the pressure recovery rate at this stage is determined mainly by the reservoir properties. In other words, water injection has little influence on the reservoir pressure at this stage.

5.2.3.2 Production Characteristics

Compared with the neighboring producers without advanced injection, the advanced-flooded producers obviously have higher outputs, which shows a certain effect of advanced injection (Fig. 5-15).

In terms of output decline, the advanced-flooded producers in southwest Wangyao seem to decline at rates similar to those nonadvanced-flooded producers in the same block. But in southwest Xinghe, the decline rate of the advanced-flooded producers is lower than that of the neighboring nonadvanced-flooded producers (Tables 5-10 and 5-11).

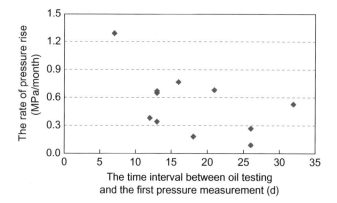

FIGURE 5-14 Relationship between the pressure rise rate and the time interval between oil testing and pressure measurement

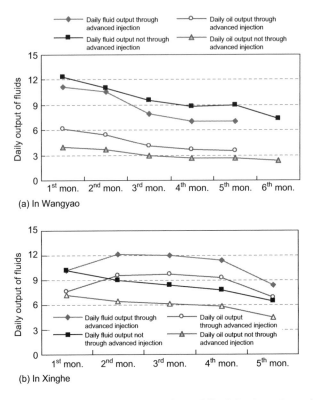

FIGURE 5-15 A comparison of outputs between advanced-flooded and nonadvanced producers.

As can been seen in the tables of the results of water injection with different timing schemes, the delayed-flooded producers show a quick decline of the initial output, which also lasts long. In spite of a modest increase one year later, the general outputs of these producers are significantly lower than those from the synchronous-flooded and advanced-flooded ones. As for the synchronous-flooded producers, the initial decline rate was obviously less than the delayed-flooded producers, and half a year after production, when these producers begin to have a response to water injection, their outputs return to higher ratios of the initial level and stay there for two years. By contrast with synchronous-flooded producers, the advanced-flooded producers have a shorter response time, of only four months, and have a noticeably higher per-well productivity, which can last longer (Table 5-12).

From the angle of different timing schemes in the same injector-producer group, advanced injection also produces a significantly better effect than synchronous injection. Take, for instance, the Wang-27-011 group in the middle-west of Wangyao. The injectors in the group began operating in September 1998, with four producers in the group put into production at the same time, namely, Wang-26-010, Wang-26-011, Wang-28-010 and

TABLE 5-10 Production Performance of Advanced-Flooded and Nonadvanced-Flooded Neighboring Producers in Southwest Wangyao

| Month | Advanced-flooded producers (22) | | | | | Nonadvanced-flooded producers (14) | | | | |
	Liquid output (m³/d)	Oil output (t/d)	Water-cut (%)	Producing liquid level (m)	Decline rate (%)	Liquid output (m³/d)	Oil output (t/d)	Water-cut (%)	Producing liquid level (m)	Decline rate (%)
1st	11.21	6.2	33.8	612		12.32	4.04	57.5	520	
2nd	10.63	5.51	37.2	864	11.2	11.08	3.67	56.6	881	9.2
3rd	7.94	4.09	39.1	1066	34	9.61	2.96	58.3	818	26.7
4th	7.13	3.63	40.1	1033	41.5	8.83	2.68	58.8	863	33.7
5th	7.04	3.56	40	1041	42.6	8.92	2.68	59.3	948	33.6

TABLE 5-11 Production Performance of Advanced-Flooded and Nonadvanced-Flooded Neighboring Producers in Southwest Xinghe

Month	Advanced-flooded producers (11)					Nonadvanced-flooded producers (10)				
	Liquid output (m³/d)	Oil output (t/d)	Water-cut (%)	Producing liquid level (m)	Decline ratio (%)	Liquid output (m³/d)	Oil output (t/d)	Water-cut (%)	Producing liquid level (m)	Decline ratio (%)
1st	10.22	7.7	10.2	695		10.17	7.24	15.4	834	
2nd	12.09	9.57	5.8	697	−24.2	9.06	6.45	14	1020	10.9
3rd	12.07	9.75	4.3	1060	−26.6	8.39	6.13	12.8	1158	15.4
4th	11.37	9.25	3.8	1103	−20.2	7.79	5.81	12	1173	19.8
5th	8.46	6.87	3.9	1126	10.8	6.4	4.56	13.5	1215	37

TABLE 5-12 Waterflood Effect at Different Timing of Water Injection in Wangyao Area of Ansai Oilfield

Injection timing	Initial output (t/d)	Before response		After response			During the stable period		Response cycle (months)
		Output at the time of response (t/d)	Ratio to initial output (%)	Maximum output after response (t/d)	Ratio to initial output (%)		Output (t/d)	Ratio to initial output (%)	
Advanced	5.31	5.01	94.40	7.18	135.20		4.66	87.80	4~5
Synchronous	4.55	4.27	93.80	5.42	119.10		3.33	73.20	5~6
Delayed < 0.5 yr	5.21	3.89	74.70	4.93	94.60		3.18	61.00	6~8
Delayed > 0.5 yr	5.15	2.53	49.1	3.51	68.2		2.93	56.9	1~5

Wang-27-012 (having encountered water breakthrough and excluded from the comparison). But the other four oil wells, i.e., Wang-26-012, Wang-27-012, Wang-28-011, and Wang-28-012, were put into production in July and August 2000, which means 23 months of advanced flooding. As a result, the daily per-well output of the four advanced-flooded wells was over 1.0 t higher than that of the synchronous-flooded four (Fig. 5-16).

5.3 DEVELOPMENT OF THE XIFENG OILFIELD

5.3.1 General Overview

Experimental advanced injection was carried out in the Xi-13 and Xi-17 well blocks in 2001. The years 2003 to 2005 saw large-scale capacity building with advanced injection, producing a total output of 110.5×10^4 t. The Chang-8 reservoirs in these blocks have a depth of 1700 to 2220 m, an initial pressure of 15.8 to 18.1 MPa, and a saturation pressure of 8.66 to 13.02 MPa. The viscosity and density of oil in place are 1.00 mPa s and 0.723 g/ml. respectively. The formation water was of $CaCl_2$ type, with a total salinity of 49.35 g/l and a pH of 6.0. The average effective thickness of the reservoir is 12.3 m, with the effective porosity averaging 10.5%, and the permeability averaging 1.27×10^{-3} μm^2.

At the end of 2007, the Xifeng Oilfield had altogether 1432 oil producers, with 1320 flowing, and 485 water injectors, with 437 at work. The average liquid output per well was 3.1 m^3/d, containing 2.3 t oil. The composite water-cut was 26.4%. The average water injected per well was 30 m^3/d. The monthly injection/production ratio was 2.32 and the cumulative injection/production ratio was 2.03. In 2007, the yearly oil output was 103.54×10^4 t. The producing rate of OIP was 0.89% and the reserve recovery was 3.17%.

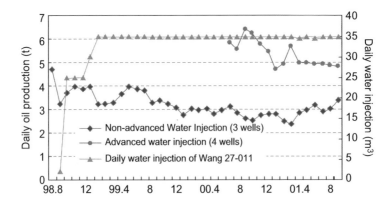

FIGURE 5-16 A comparison between advanced injection and synchronous injection in the Wang-27-011 well group.

5.3.2 Technological Policies

Between 2003 and 2005, the producing zones in Xifeng underwent 4 to 6 months of advanced flooding, during which the average water injected per well was 31 m^3/d.

5.3.2.1 The Best Injection Timing

It can be seen from the correlation curves of production performance in central Baima (Fig. 5-17) and the effect of different timing schemes of advanced injection (Table 5-13) that oil wells with six months of advanced water injection show a low decline rate at the initial stage of production and a short response time, and that their outputs in the response period are obviously higher than those from the producers with two to three months or those with more than six months of advanced injection, thus bringing about a better effect. Compared with the producers with four to five months of advanced injection, those with two to three months of advanced injection, show a slightly longer response time and a moderately higher daily output both at the response stage and at the stable stage. With the overall economic efficiency in mind, we conclude that four to six months of advanced water injection is reasonable in the development of the ultralow-permeability Chang-8 layers of the Xifeng Oilfield, which is consistent with our theoretical calculations before.

5.3.2.2 Selection of Reasonable Injection Parameters

Theoretical calculations indicate that the maximum water injection intensity should be kept at 2.51 m^3/d m in central Baima, at a daily rate of about 30 m^3.

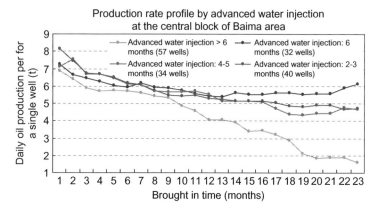

FIGURE 5-17 A comparison of output changes in producers with different timing schemes of advanced injection in the central Baima block in 2002 and 2003.

According to the actual effect of advanced injection in the block, an injection intensity exceeding $2.0\,\mathrm{m^3/d\,m}$ may produce early water breakthrough and quick rise of water-cut after oil wells are put into production (Fig. 5-18), which agrees with our theoretical calculation. In order to avoid intralayer fingering and early water breakthrough, the intensity at the initial stage should therefore be kept at less than $2.0\,\mathrm{m^3/d\,m}$. In conclusion, the

TABLE 5-13 Effect of Advanced Injection in Central Baima of the Xifeng Oilfield

Injection timing	Before response (%)			After response (%)		During the stable period (%)		Response cycle (months)
6 months in advance	6.40	5.69	88.91	7.25	113.28	5.62	87.81	2~4
6 months in advance	6.79	5.92	87.19	7.52	110.75	5.68	83.65	3~5
4–5 months in advance	7.11	5.68	78.89	6.42	90.30	5.65	79.47	4~6
2–3 months in advance	7.45	5.48	73.56	6.13	82.28	5.43	72.89	4~6

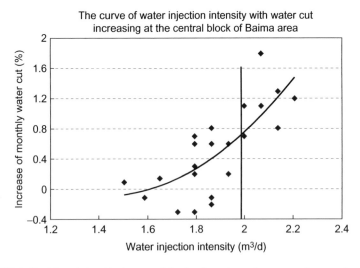

FIGURE 5-18 The relationship between injection intensity and water-cut increase in central Baima.

daily injection rate per well and the injection intensity should be kept at 20 to 25 m^3 and 1.6 to 2.0 m^3/d m, respectively.

5.3.3 Results of Advanced Injection

5.3.3.1 Characteristics of Pressure Changes

Thanks to overall advanced injection, the reservoir energy in the Xifeng Oilfield was kept at a high level, with pressure changes presenting two features.

The advanced-flooded producers contained relatively high initial pressures, which were maintained at a high level.

- As the advanced-flooded producers in central Baima were put into production during 2003 and 2004, their formation pressure exceeded 20 MPa, 110% of the original. Similarly, the advanced-flooded producers in Dongzhi contained a formation pressure over 16.6 MPa when put into production in 2003 and 2004, 110.5% of the original (Figs. 5-19 and 5-20).
- The reservoir pressure in the block rose year by year, with the rate of rise slowing down.
- By overall advanced injection in Baima and Dongzhi, the reservoir pressure in both blocks was kept at a relatively high level at the initial stage of production. In central Baima, the initial pressure reached up to 16.8 MPa, 92.9% of the original and rising every year. In Dongzhi, the initial pressure reached up to 14.8 MPa, 96.7% of the original and also rising every year.

5.3.3.2 Characteristics of Output and Its Decline

Compared with 75 synchronous-flooded wells and 38 wells developed by natural energy out of the pilot blocks, the advanced-flooded producers

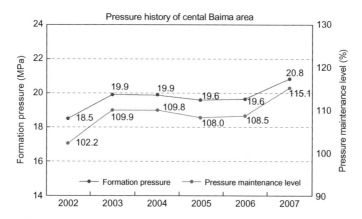

FIGURE 5-19   Reservoir pressure changes in central Baima.

presented a higher initial productivity (in the first three months), reaching up to 6.8 t/d per well, while the former two kinds of producers produced only an average of 5.35 t/d and 3.02 t/d, respectively.

The advanced-flooded producers operated at a high productivity, which declined slowly and was stabilized for a long time, with a decline of only 9.6% at the initial stage of production (Fig. 5-21). The wells operating through synchronous flooding or natural energy, on the contrary, showed a higher rate of decline at the initial stage of production, reaching up to 35.8% and 44.16%, respectively.

In addition, the productivity of advanced-flooded producers, with a shorter period of decline, turned up from the response time of four to six months.

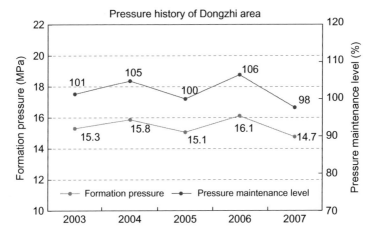

FIGURE 5-20   Reservoir pressure changes in Dongzhi.

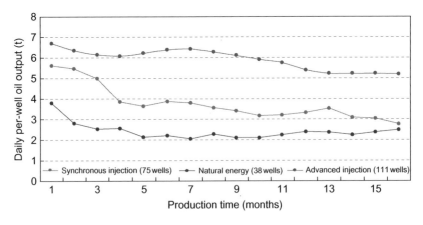

FIGURE 5-21   A comparison of oil production history in Baima, developed through advanced injection, natural energy, and synchronous injection.

On the other hand, the period of output decline in wells operating through synchronous flooding or natural energy lasted as long as 6 to 7 months.

5.3.3.3 Characteristics of Producer Responses

Overall advanced injection carried out in Xifeng shortened the response time in the producers whose productivity rose while the water-cut remained stable afterwards. There are two characteristics of producer response.

After the oil wells began to respond to injection, both their liquid and oil outputs began to go up while their water-cut remained stable (Table 5-14).

Xi-25-21 is a typical advanced-flooded producer, whose productivity in the first three months was 6.82 t/d on average. Four months later, when water injection became effective, its productivity rose from an average of 6.72 t/d in the first four months to 9.51% afterwards, with the water-cut stable. And such stable production lasted as long as 13 months (Fig. 5-22). Another typical advanced-flooded producer is Xi-25−28, whose productivity in the first three months was 7.77 t/d on average. Five months later, when water injection became effective, its productivity rose from an average of 5.48 t/d in the first 5 months to 6.35% afterwards, with the water-cut stable. Such stable production lasted as long as 8 months (Fig. 5-23).

Advanced injection produced a short response time in oil wells, whose outputs tended to go up considerably afterwards.

From the production history of advanced injection, synchronous injection and delayed injection in central Baima, it can be seen that the advanced-flooded oil wells showed the shortest response time, usually 3 to 6 months, with a following output of up to 98.45% of the initial. On the contrary, the response time of oil wells developed by synchronous injection and natural energy was relatively longer, 5 to 7 months and 9 to 12 months, respectively, and the following outputs reached a relatively lower percentage of their initials, 72.52% and 78.48%, respectively (Fig. 5-24 and Table 5-15).

5.4 DEVELOPMENT OF THE NANLIANG OILFIELD

5.4.1 General Overview

Exploratory drilling began in the Nanliang Oilfield in 1970. In 2001, pilot development through advanced injection was initiated in the Wu-10 and Wu-11 blocks in western Nanliang. During 2002 and 2006, the oilfield continued rolling development, which produced a total output of 26.0×10^4 t. The Chang-4 + 5 reservoirs in these zones have a depth between 1848 and 1876 m, an average thickness of 14.2 m, an effective porosity averaging 14.2%, and a permeability averaging 0.49×10^{-3} μm^2. In addition, the reservoirs have an initial pressure of 12.61 to 15.74 MPa, a pressure coefficient of 0.84 to 1.10, and a saturation pressure of 0.69 to 7.13 MPa. The oil in place has a viscosity of 1.05 to 1.45 mPa s and a density of 0.84 g/ml.

TABLE 5-14 Effect of Advanced Injection in Central Baima of Xifeng

Before response				After response				During the stable period			
Daily liquid production (m³)	Daily oil production (t)	Water cut (%)	Dynamic liquid level (m)	Daily liquid production (m³)	Daily oil production (t)	Water cut (%)	Dynamic liquid level (m)	Daily liquid production (m³)	Daily oil production (t)	Water cut (%)	Dynamic liquid level (m)
6.10	5.70	6.62	1221.00	7.20	6.80	5.56	1204.00	6.01	5.60	7.05	1208.00

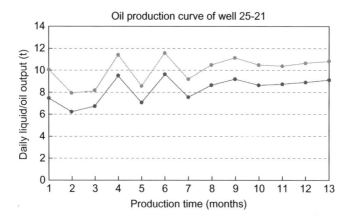

FIGURE 5-22   Production curves of the typical response well Xi-25-21.

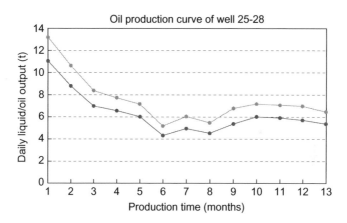

FIGURE 5-23   Production curves of the typical response well Xi-25-28.

At the end of 2007, there were 366 production wells in the Oilfield, 332 in operation, with the liquid output averaging 3.9 m³/d per well, the oil output averaging 2.5 t/d per well, and the composite water-cut at 36.7%. At the same time, there were 147 water injectors, 145 wells flowing, with the daily injection rate averaging 38 m³ per well, a monthly injection/production ratio of 3.53, and a cumulative injection/production ratio of 2.8. In 2007, the yearly oil output reached 26.05×10^4 t, with a recovery rate of 1.1% and a recovery factor of 5.37%.

5.4.2 Technological Policies

The pilot development of Wu-10 and Wu-11 began in August 2001, with 30,000-ton-water advanced injection, at a rate of 38 m³/d per well. The

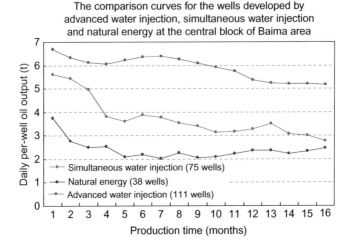

FIGURE 5-24 A comparison of output decline and oil wells developed by advanced injection in Xifeng during 2002 and 2003.

corresponding oil wells were put into production in October and November that year, 60 to 101 days after the injection began. Since 2002, thicker zones with relatively higher permeabilities have been picked up for capacity building through advanced injection.

5.4.2.1 Reasonable Timing for Water Injection

Fig. 5-25 is a statistical analysis of the relationship between the per-well productivity and the water-cut in the oil producers with different timing schemes of advanced injection in western Nanliang.

By comparing the development effects of advanced injection with different timing schemes, we can see that the per-well outputs of oil producers with six to nine months of advanced injection are the highest, and their water-cuts keep stable. Thus, we conclude that the best timing for water injection in this block is 6 to 9 months in advance.

5.4.2.2 Reasonable Water Injection Intensity

It can be seen from the injection parameters in western Nanliang that the initial injection rate lies between 30 and 40 m^3 a day, the injection intensity between 3.0 and 4.0 m^3/d m. It turns out that the producers with 6 to 9 months of advanced injection have the highest initial per-well productivity, which declines at a low rate. In addition, these producers show a higher average daily output and a stable water-cut (Figs. 5-26 and 5-27).

Statistics of the relations between injection intensity on the one hand and per-well productivity and water-cut rise on the other show that, when the

TABLE 5-15 Effect of Advanced, Synchronous, and Delayed Injection in Central Baima of Xifeng

Injection timing	Initial output (t/d)	Before response		After response			During stable period		Response cycle (months)
		Output (t/d)	Ratio to initial output (%)	Maximum output (t/d)	Ratio to initial output (%)		Output (t/d)	Ratio to initial output (%)	
Advanced	6.94	5.69	82.05	6.83	98.45		5.60	80.65	3~6
Synchronous	5.35	3.61	67.48	3.88	72.52		3.40	63.55	5~8
Delayed	3.02	2.11	69.87	2.37	78.48		2.03	67.22	9~12

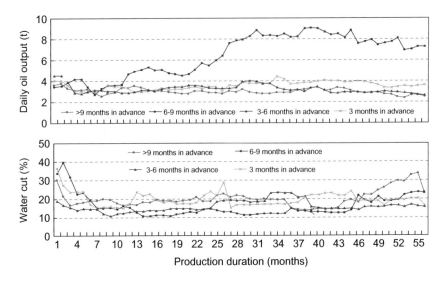

FIGURE 5-25 A comparison of production history of well groups with different timing schemes of advanced injection in western Nanliang.

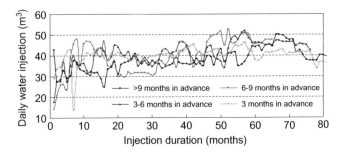

FIGURE 5-26 Curves of daily outputs with different timing schemes of advanced injection in western Nanliang.

injection intensity is kept at 3.0 to 4.0 m³/d m, the per-well productivity remains high while the water-cut rise slows down. But when the injection intensity exceeds 4.0 m³/d m, the water-cut rise may be quickened significantly. Therefore, the injection intensity should be held between 3.0 and 4.0 m³/d m, and the daily injection rate should be held between 30 and 40 m³ (Figs. 5-28 and 5-29).

These statistics and our experience in western Nanliang indicate that an injection intensity between 3.0 and 4.0 m³/d m would not cause intralayer water rush or fingering and premature water breakthrough into the producer, but would lead to relatively high per-well productivity and stable water-cut.

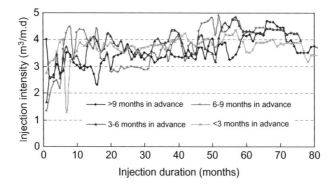

FIGURE 5-27 Curves of injection intensities with different timing schemes of advanced injection in western Nanliang.

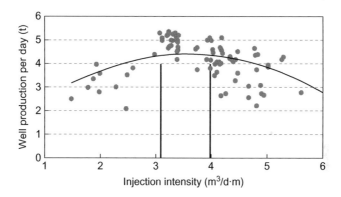

FIGURE 5-28 The relationship between injection intensity and well productivity in western Nanliang.

In conclusion, we conclude that the advanced water injection in Nanliang should be kept at a rate of 40–50 m³/d per well and an intensity between 3.0 and 4.0 m³/d m.

5.4.3 Results of Advanced Water Injection

5.4.3.1 Characteristics of Pressure Changes

The original formation pressure in the experimental zone was 13.7 MPa. Advanced injection in the zone lasted two months, with a total amount of water injected reaching 7671 m³. The curve of pressure changes through numerical simulation shows that, as the injected volume increased, the pressure began to recover and, by the end of water injection, the average pressure reached 13.80 MPa, 100.7% of the original (13.7 MPa), which indicates that advanced injection became effective. In 2002, advanced injection was carried out on a larger scale across the whole capacity building area.

A comparison of pressure changes in advanced-flooded well groups with those in nonadvanced-flooded well groups in recent years shows that the pressure in the former was kept at a higher level and more stable than that in the latter. In the nonadvanced-flooded wells that were put into production between 2003 and 2004, the pressure increased at the initial stage but declined later, whereas the pressure in the advanced-flooded zones was kept at a higher level all the time (Fig. 5-30). It can therefore be concluded that advanced injection in ultralow-permeability reservoirs can help increase per-well output through pressure maintenance on the one hand, and on the other

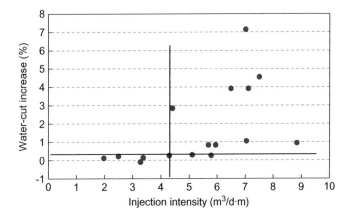

FIGURE 5-29 The relationship between injection intensity and water-cut increase in western Nanliang.

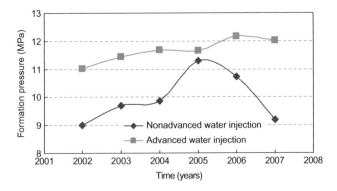

FIGURE 5-30 Pressure history of advanced-flooded zones and nonadvanced-flooded zones in western Nanliang.

hand, can help reduce the loss of reservoir productivity caused by pressure sensitivity, which may damage porosity and permeability.

5.4.3.2 Characteristics of Production Decline

The data of daily oil and liquid outputs show that the maximum output from the experimental well group reached 54.07 t/d (December 2001) and the minimum was 36.70 t/d (April 2002), at a monthly decline rate of 8.03%, lower than the same type of oilfields not using advanced injection. Such statistics suggest that advanced injection can help slow down the output decline at the initial stage of oilfield development.

A comparison of productivity decline between advanced-flooded zones and nonadvanced-flooded zones (Fig. 5-31) shows that the former experience a shorter period and a lower rate of decline (15%) than the latter (33.4%) at the initial stage of production. It can be concluded that advanced injection in the ultralow-permeability reservoirs succeeds in maintaining the reservoir energy, alleviating its stress sensitivity and reducing the decline of well productivity.

A comparison of the response time between advanced-flooded zones and nonadvanced-flooded zones shows that the former begin to respond to water injection four months after commissioning, with the per-well productivity increasing, whereas in the latter, responses are not seen until eight months later, when the productivity stops declining and begins to rise. It can be concluded then that advanced injection shortens the decline period of the initial productivity and reduces its decline rate as well.

5.4.3.3 Characteristics of Water Breakthrough and Producer Response

The curves of water breakthrough and producer response of advanced and nonadvanced injection (Fig. 5-31) show that the advanced-flooded well groups respond four months after commissioning, with the per-well

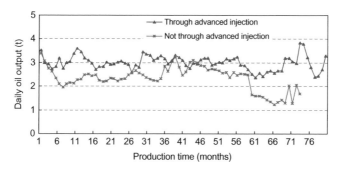

FIGURE 5-31 A comparison of per-well productivity between advanced-flooded zones and nonadvanced-flooded zones in western Nanliang.

productivity increasing and the water-cut stable, whereas the nonadvanced groups present a response eight months after commissioning. The production dynamics before and after response in the two groups are presented in Table 5-16.

It can be seen from the table that the advanced-flooded groups have a higher initial per-well productivity of 2.6 t/d, with a water-cut of 18% and a producing fluid level of 1450 m. After response, their per-well productivity increases to 3.0 t/d, with the producing fluid level rising to 1400 m and the water-cut remaining stable. At the stable stage, both their per-well productivity and water-cut are stabilized, while the producing fluid level falls a little. On the other hand, the nonadvanced-flooded groups show a relatively low initial productivity of 2.5 t/d, with a water-cut of 19%. After response, their daily productivity rises to 3.0 t per well. And in the stable period, the per-well productivity remains at about 2.7 t/d while the producing fluid level falls down to 1620 m.

In general, the well groups with advanced injection show a higher initial output and shorter response time. After response, the productivity of these wells, with stable water-cuts, can increase considerably and stay at a high level. In a word, the results of advanced injection are obviously better than those of synchronous injection or delayed injection in oil production.

CONCLUSIONS

Fifty years have passed since the Changqing Oilfield Company began its petroleum exploration and development in the Ordos Basin. Great breakthroughs have been made between the late 1980s and the beginning of this new century, especially in recent years, through the discovery and effective development of quite a few large oilfields such as Ansai, Jing'an, Xifeng, Shuijing, and Jiyuan, making Changqing one of the largest oil producers in China, with an annual output of 18.5 million tons. This is not only a process of deepening our understanding of the low- or ultralow-permeability oilfields, but also a process of continual innovation, improvement, and perfection of technologies for the exploration and development of low-permeability reservoirs. Advanced water injection, developed from theoretical studies, lab tests, and field experiments through to wide application, has become one of the core technologies for the development of ultralow-permeability reservoirs. Meanwhile, the integration of fine reservoir description, well-pattern optimization, advanced water injection, and reservoir fracturing enables the Changqing Oilfield Company to explore and develop the ultralow-permeability reservoirs with high efficiency, and therefore, to become a world leader in this field.

While the Ordos Basin is one of China's major energy bases, the Changqing Oilfield Company happens to be the main explorer and developer in the Basin. As a major oil producer in China, we will, taking into consideration the country's economic boom and increasing demand for oil and gas

TABLE 5-16 Production Dynamics of Advanced-Flooded and Nonadvanced-Flooded Wells in Western Nanliang before and after Response

Injection mode	Before response				After response				Stable production period			
	Daily fluid output (m^3)	Daily oil output (t)	Water cut (%)	Producing fluid level (m)	Daily fluid output (m^3)	Daily oil output (t)	Water cut (%)	Producing fluid level (m)	Daily fluid output (m^3)	Daily oil output (t)	Water cut (%)	Producing fluid level (m)
Advanced water injection	3.5	2.6	18	1450	3.9	3	18	1400	4	3.1	20	1550
非超前注水	2.5	2	19	1450	3	2.3	23	1500	3.9	2.7	27	1620

and on the basis of our experience in low-permeability oilfield development, continue to create room for the development of such fields, taking a lead and tackling new discoveries in this field so that we can contribute more to the Chinese oil industry.

This book was written by Professor RAN Xinquan with the assistance of CHENG Qigui, QU Xuefeng, LIU Lili, LI Xianwen, and other experts, also from the Changqing Oilfield Company. The author would also like to extend his thanks to Professor CHENG Linsong of the China University of Petroleum (Beijing), Professor TANG Hai of the Southwest Petroleum University, and Professor ZHANG Chunsheng of the Yangtze University, whose advice played an important role in the accomplishment of this book.

Index

Note: Page number followed by "*f*" and "*t*" refer to figures and tables, respectively.

Printed and bound by CPI Group (UK) Ltd, Croydon, CR0 4YY

13/10/2024

01773514-0001